Mammalian Cardiovascular System Simulation

A Catastrophe Theoretic Approach with the Matching Simulation Method

J L Janssens & W Kuyk
Department of Mathematics & Information Science
Laboratory for Chronobiology
Universiteit Antwerpen
Groenenborgerlaan 171
B-2020 Antwerpen 2 Belgium

Wuerz Publishing Ltd
Winnipeg, Canada

Wuerz Publishing Ltd
Winnipeg, Canada

Mammalian Cardiovascular System Simulation:
A Catastrophe Theoretic Approach with the Matching Simulation Method

Janssens, Jan L, and Kuyk, Willem

ISBN 0-920063-48-9

Printed in Canada

Biological Systems
A Book Series

Wuerz Publishing Ltd
895 McMillan Avenue
Winnipeg MB R3M 0T2
Canada

phone (204) 453 7429
fax (204) 453 6598

Preface

In this essay we simulate the oscillating dynamic of the mammalian cardiovascular system from an electric model of the vascular system joined to a 'catastrophe machine' that simulates the ventricular double-acting pump. We have construed the machine in such a way as to have it act on and respond to the vascular model in this sense: the inotropic changes in the machine (beat rate, state of contractility, etc.) and changes in the vascular system (blood volume, aortic impedance, pressures, etc.) are dynamically coupled as in the real cardiovascular system. As a result, the simulation is a mathematical-physical model consisting of eight integral-differential equations whose periodic solutions yield the usual time graphs of the mammalian heart. Obviously, the mathematical model is not purely deduced from basic biophysical principles, mainly because it is not known how even the quantifiable and measurable heart parameters fit together into one deductive model –witness the present ignorance of the fundamental biochemical and physiological laws governing muscle control–, and partly since there are no known ways to parametrize some of the physiological parameters that control the heart pump (e.g., contractility). The method that we apply we call the "matching-simulation method". It consists in finding a refining series of models that simulate the real behaviour of the heart, in such a way that all the main parameters that are known to control the cardiovascular system are integrated judiciously into the model so as to play the role as they do in fact play. Our series consists of a fixed one for the vascular system, placing it between the model of Sagawa, Suga and Nakayama (1983) on the one hand and that of Rideout and Snyder (1969) on the other; and two approximating catastrophe models for the ventricle, namely, a three-dimensional cusp model that refines to a five-dimensional model. The latter is the so-called "butterfly model; it has four control parameters (a parameter called 'dynamic parameter", the pre- and afterload, and ventricular contractility), and one state parameter (the ventricular pressure, which is the state parameter throughout).

"Matching" in this case means a series of attempte to identify the known cardiovascular control parameters with control parameters of the mathematical cusp- and butterfly model, and then "simulating" the operation of the heart. Comparison between the recorded graphs in the literature in each case decided whether or not the match was right. Dynamic changes were taken into account since we required that changes of the parameters in the model should run 'parallel' to changes of the parameters of the real heart pump. Note that within every chapter, the scale of the state and control parameters is consistent, without converting them to conform to the values that belong to any particular double-acting pump. However, in applications one may rescale or gauge them to suit any desired particular case. The models may be useful in the construction of artificial heart pumps, and may be used in teaching situations as an alternative to live-animal experiments.

The monograph has nine chapters the contents of which may be circumscribed as follows.

In Chapter 1 we recall briefly the standard physiological knowledge of the cardiovascular system, explaining the main cardiac parameters: frequency, preload, afterload and myocardial contractility. Readers versed in cardiology might skip this chapter.

Chapter 2 highlights the general characteristics of Catastrophe Theory applied to the cusp catastrophe. The latter is a catastrophe type whose full three-dimensional unfolding has one state and two control parameters, a two-dimensional bifurcation set and two catastrophic jumps. It illustrates hysteresis and divergence. The chapter is basic to all the catastrophe terminology that is to follow.

In Chapter 3 we exploit the observation by Zeeman that a van der Pol oscillator with large damping may be interpreted as a cusp catastrophe of one state parameter, with one fixed and one dynamic parameter as control parameters. The state and the dynamic control parameters are linked by a set of two first-order differential equations, which are called the 'slow' and the 'fast equation'. The fast equation sets the time derivative of the state parameter $\dot{x}$ equal to a polynomial of third degree in x. The two lowest coefficients of this polynomial are a_2 and a_1 which are equal to the fixed and the dynamic control parameter, respectively, and the term in x^2 is zero. The higher odd-order polynomials with only two non-zero coefficients for the lowest order terms, lead also to the cusp catastrophe type since there exists a diffeomorphism between the surfaces defined by the polynomials. Reshaping a system to a minimal polynomial by way of a diffeomorphism is a common procedure in Catastrophe Theory. Unfortunately, diffeomorphisms do not necessarily preserve the physical-dimensional relations between the parameters. A requisite for this chapter is some elementary knowledge on differential equation.

In Chapter 4 we recall the known representations of ventricular events. The latter are the two pressure jumps, at almost constant volume; the fast and slow ejection during systole; the ventricular filling during diastole and an additional filling during atrial systole. These events are discussed on the basis of what we call the 'recorded graphs'; i.e., the graphs that are found in the standard literature and that have been derived and sometimes guessed from in-vivo and in-vitro experiments. In this essay we focus on the ventricular activity and omit the atrial activity. Also, the pulmonary circulation, which is not fundamentally different from the systemic circulation is being omitted. The contents are standard knowledge to be found in most medical textbooks.

In Chapter 5 we discuss into detail the classical simulation of the cardiovascular system as presented by Rideout and Snyder. This simulation is based on a set of integral-differential equations for the vascular system and a time-varying compliance for the left ventricle. Using the computer program TUTSIM, we numerically solved the set of integral equations. The resulting 'simulated graphs' are compared with the recorded graphs from Chapter 4. In general the simulated and recorded graphs are very much alike, but they differ at the place in which atrial activity is significant. A serious drawback for this model is that the artificial time-varying compliance cannot be made to match with the earlier mentioned cardiac parameters.

In Chapter 6 we describe the matching-simulation method, and match the full three-dimensional cusp model to the ventricle. The state and two control parameters of the cusp model are the ventricular pressure and volume and a dynamic parameter. The latter parameter forces the catastrophic jumps to take place and determines the period of the system, whereas the volume, which quantizes the preload, is the splitting parameter. The simulation of this model shows that this match models several features of the cardiovascular system but not contractility and afterload. To obtain a better approximation we increase the number of control parameters from two to four and match the ventricle to the butterfly model.

In Chapter 7 we describe the full five-dimensional unfolding of the butterfly catastrophe, computing features that are not to be found in the literature on catastrophes. We study the shape in (a_1,x)-space of the catastrophe surface for different fixed values of the other control parameters. We express the distance between the extremal intersections of the x-axis and the butterfly hypersurface as a function of the values of the control parameters. This function is required for subsequent chapters.

In Chapter 8 we apply the matching simulation method to show that near the origin, the full five-dimensional unfolding of the butterfly catastrophe is needed to model the ventricle. In addition to the parameters that this model has in common with the cusp model, and have already been matched, the two new control parameters are identified with the aortic pressure and myocardial contractility. It is easily seen that the aortic pressure quantifies the afterload. The graphs obtained from this model bear a very good resemblance to the recorded graphs in Chapter 4.

Finally, in Chapter 9 we show that the butterfly model not only models the quasistatic behaviour of the ventricle but also that it models the dynamic auto-regulation of the human cardiac suction pump. More precisely, we show how the Frank and Starling laws for the heart are contained in the butterfly model. In fact, as shown in the ventricular volume-ventricular pressure plane, sections of the five-dimensional behaviour of the heart that depict the Frank-Starling law, and show them to accord with the well-known Sonnenblick graphs.

We wish to thank the following people for several useful discussions of our work, and for reading the manuscript and proposing improvements of contents and text: Dr J.A. Metz (Leiden University, The Netherlands), Dr E. Raman (Ruca, Antwerp) and Dr S. Sys (Ruca, Antwerp).

J.L. Janssens and W. Kuyk
Department of Mathematics
University of Antwerp
Rijksuniversitair Centrum
Groenenborgerlaan 171
2020 Antwerp
Belgium

List of Abbreviations and pages of first appearance

AV, atrio-ventricular (node), 2
B, bifurcation set (of manifold), 13
C, compliance, 8
E, modulus of elasticity, 8
EDV, end diastolic volume, 3
ESV, end systolic volume, 26
F, arterial flow, 8
h, arterial wall-thickness, 8
IVC, inferior vena cava, 30
K, damping constant (of oscillator), 18
L, inertance, 8
l, length of artery, 8
LA, left atrium, 30
LVP, left ventricular pressure 26
M, catastrophe manifold, 13
P, ventricular pressure, *passim*
PSP, peak systolic volume, 48
R, resistor, 8
r, arterial radius, 8
RA, right atrium, 30
S , time-scaling factor, 84
SA, sino-atrial (node), 2
SV, stroke volume, 4
SVC, superior vena cava, 30
T, oscillation period, 84
U-line, locus of maximal ventricular pressure, 98
V, ventricular volume, *passim*

ϵ, damping term, 17
μ, blood viscosity, 8
ρ, blood density, 8
σ, Poisson rate, 8
τ, relaxation time (of oscillator), 17
ω, pseudo-relaxation, 17

Chapter 1 The Cardiovascular System

1.1 Introduction

The heart and two vascular systems build the mammalian circulation. An adult mammalian heart consists of two fluid pumps that anatomically are side by side, but functionally are connected in series. The right heart supplies blood to the lungs through the pulmonary circulation. The left heart supplies the remainder of the body via the systemic circulation before returning the blood to the right heart. The left heart consists of two chambers (the left *atrium* and the left *ventricle*) which are separated by an atrio-ventricular valve (AV). The aortic valve separates the left ventricle from the systemic system. The right heart consists of the right *atrium* and the right *ventricle* which are also separated by an atrio-ventricular valve. The pulmonary valve separates the right ventricle from the pulmonary artery.

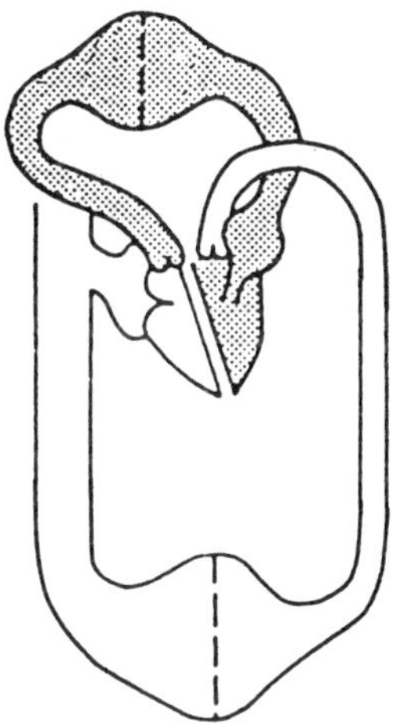

Fig. 1.1. Scheme of the circulation system The shaded area indicates the pulmonary system (from Milnor in Brobeck (1979))

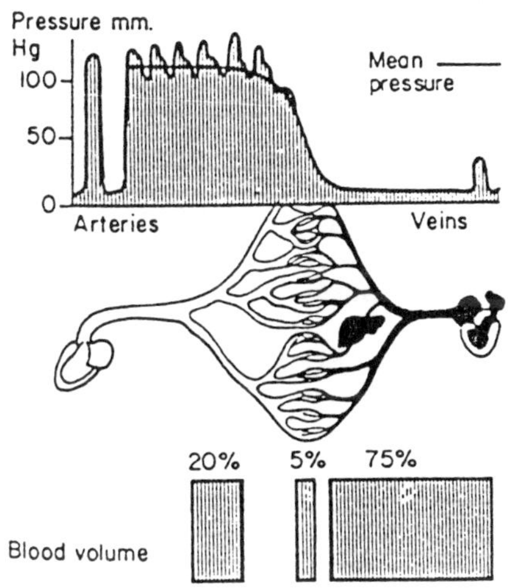

Fig. 1.2. Volumes and pressure in the systemic circulation (from Brobeck (1979))

The vascular system can be divided into three parts, namely

1 aorta and arteries, as a distribution system,

2 the microcirculation, the diffusion and filtration system, and

3 the veins, as a collection system.

Normally, the global mean flow through the pulmonary and systemic systems should be the same. The pressure built up in the ventricle causes fluid flow through the vascular system, by having the ventricle contract on its contents, ejecting the blood through the

AV-valves. Hence, there is a pressure drop downstream from the aorta and arteries, where the pressure is high and pulsing, through the microcirculation, where the drop is highest, and through the veins, where the pressure is low and relatively constant. For adult humans, the total mass of blood is about $5 \times 10^{-3} m^3 (5l)$. When the blood volume is increased, as during transfusion, only 1%. of the supplied volume remains in the high pressure system. Thus the veins serve as a reservoir. As a collecting system it contains up to 75%. of the total blood volume whereas the microcirculation contains only 5% and the distributing system the remaining 20%. For a more detailed description of the cardiovascular system we refer to (Brobeck 1979) and (Ganong 1973)

1.2 The cardiac cycle

Each AV-valve allows a pressure difference between the veins and ventricles adjacent to it. When the ventricles contract, the ventricular pressures raise, so that the atrio-ventricular valves close. As the contractions go on in closed chambers, the ventricular pressures increase suddenly, while the ventricular volumes stay constant. This phase is called *isometric contraction.*

Then, after the aortic and pulmonary valves are forced open and blood is ejected into the aortic and pulmonary arteries, the ventricular volumes decrease. This phase is called the *ejection phase.* During this phase the aortic (and pulmonary) pressures start to rise and may exceed the ventricular pressures. The ejection then decreases, and the ventricular pressures decrease. Also the heart muscles start to relax. When the aortic and pulmonary valves close, the relaxation of the heart muscle continues, causing a steep fall in ventricular pressure. This phase is called the *isometric relaxation.* The time elapsed between the closure of the atrio-ventricular valves and the closure of the aortic an pulmonary valves is called *(ventricular) systole.* When the pressure level in one of the ventricles is lower than the pressure in the corresponding atrium, blood flows into the ventricle. The rapid filling of the ventricle causes a slight pressure rise in the ventricle. The filling slows down when the pressure in ventricle and atrium equalize. During the atrial contraction also fluid is pumped into the ventricle, although the increase in blood volume during this phase is much less than during the preceding phase, and the cycle can restart again. The time elapsed between the closure of the aortic and pulmonary valves, and the closure of the atrio-ventricular valve, is called *(ventricular) diastole.* Figure 1.3 sums up the main events of the cardiac cycle.

1.3 Control of the heart rate

1.3.1 Intrinsic control

The control of the circulation is highly intrinsic. Even when the heart is denervated, it remains pumping. This is due to the autostimulation of the heart muscle (cells). All the heart cells of the atrium are able to generate stimuli. But some special cells located in the sino-atrial-node (SA-node) in the right atrial wall are specialized for this task, and are called *pacemakers.* From the SA-node the heart muscle itself conducts the electrical stimulus over the atria. On its travel, this stimulus triggers the contraction of the ventricular muscle fibres. The junction between the atria and the ventricles does not conduct stimuli. The atrio-ventricular node (AV-node) is the most likely site at which

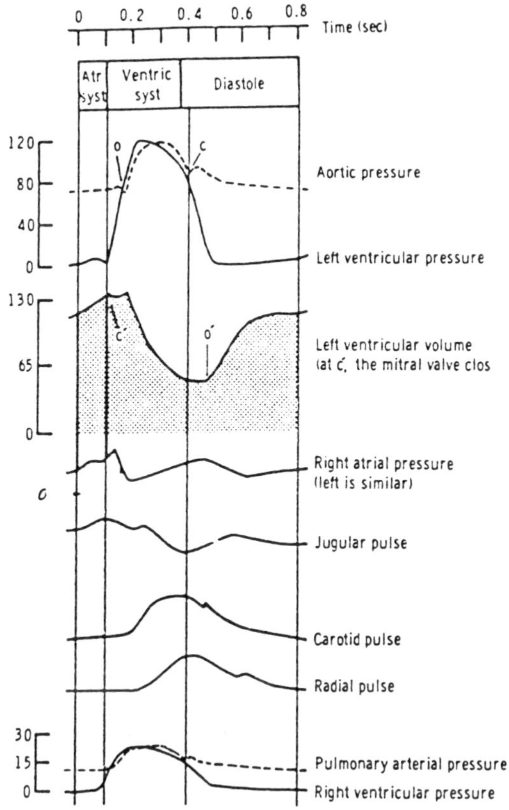

Fig. 1.3. Events in the cardiac cycle (from Ganong (1973))

the stimulus can pass from the left atrium to the ventricles. From the AV-node the Purkinje fibres conduct the stimulus very quickly over the ventricles. After stimulation the muscle fibre has an absolute refractory period of about 150 ms, during which the fibre cannot be stimulated again. During the following relative refractory period of 50 ms the muscle fibres return to their normal relaxed situation. The autostimulation of the atrial cells and of some cells in the ventricle has the advantage that, if the SA-node ceases to generate stimuli, then the other cells can take over its function. However, these cells may generate stimuli which are out of phase even with the normally functioning SA-node, so that fibres may contract in a random way, blocking the normal functioning of the heart (ectopic foci). However, in normal situations the absolute refractory period is long enough so that the stimulus dies out when all the heart cells have contracted.

Another kind of intrinsic control is known as the "Frank-Starling Law". This law states that the performance of the heart increases with any increase in the end diastolic volume EDV. For example, a higher aortic pressure decreases the outflow during systole. The rest volume in the heart remains larger and so does the next EDV. This one causes in turn a higher pressure during systole, so that the cardiac output returns to the original value after a few beats.

1.3.2 Extrinsic control

Extrinsic control of the heart rate is also possible. Transmitters, nerve impulses, and humoral agents affect the Ca^{++}-permeability properties of the heart cell membranes, so that the firing rate of pacemaker cells increase or decrease. They influence the pulsing rate, the contractility of the heart fibres, the resistance of the arterioles (small arteries) and the blood volume in the circulation system. For instance, emotions and even intended activities do change the contractility and the heart rate.

1.4 Control of cardiac output

The amount of blood pumped out of each ventricle per beat, the stroke volume (SV), is about 80 ml in a resting man in the supine position. The output of the heart per time unit is the cardiac output, and is equal to the SV times the heart rate. In a resting, supine male, it averages about 5.5 l/min = (70 beats/min∗80 ml/beat). / Variations in the cardiac output can be produced by changes in the cardiac rate or stroke volume. As discussed earlier, the cardiac rate is controlled via the cardiac innervation. The stroke volume is controlled by the EDV and by myocardial contractility, aortic pressure, etc.

1.4.1 The length-tension relationship in the cardiac muscle

The relationship between the tension and length of the cardiac muscle has been investigated by Starling. He stated as early as 1914 that the "energy of contraction is proportional to the initial length of the cardiac muscle fibre". This statement has come to be known as the "Frank-Starling Law". Different forms of the "Frank-Starling Law" are known. The initial length of the muscle fibres determine the tension in the tissue of the heart at the end of diastole. The force on the cardiac wall which causes the same tension is called the *preload*. As the EDV is proportional to the length of the cardiac muscle fibre, one obtains a relation between the ventricular performance and the EDV. This experimental, nonlinear, relationship is shown in Figure 1.5 together with a summary of the major factors affecting the EDV. Although the relationship is not linear, a linear approximation may be acceptable if the variations in the EDV are small. Under conditions which affect the EDV in a more serious way, such as excitement, eating, exercise or high environmental temperature, the relationship between ventricular performance and EDV is no longer linear. The preload and the afterload, which will be defined in 1.4.2, are very important cardiac parameters.

Another relationship which is often used is the *resting length-tension curve*. When the length of a resting heart muscle is increased by increasing the preload, then the tension rises with each increment of the length. The obtained relation between length and tension is called the resting length-tension relation. This relation is not altered appreciably by the other control parameters (from [Brobeck(1979)]). Its shape is an increasing function of the muscle length, and becomes steeper as the length increases.

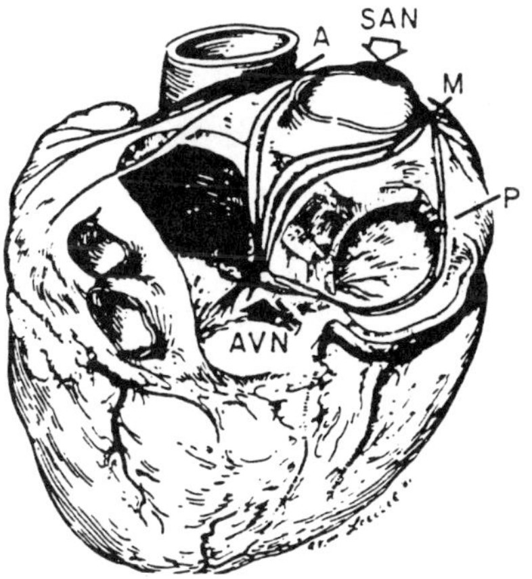

Fig. 1.4. Diagram of the stimulation nodes and conducting fibres on the heart (from James in Ganong (1973))

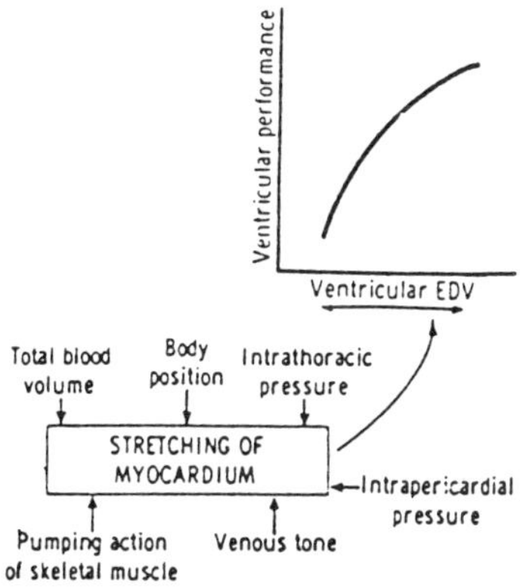

Fig. 1.5. Relation between ventricular EDV and ventricular performance (from Braunwald in Ganong (1973))

1.4.2 Changes in vascular impedance

The vascular impedance is the global resistance of the arteries and veins. The effects of changes in vascular impedance were studied with a heart-lung preparation. Such a preparation is shown in Figure 1.6; the heart and lungs of an experimental animal, usually a mongrel dog, are cannulated in such a way that the blood flows from the aorta through a system of tubings and reservoirs to the right atrium, and from there through the animal's heart and lungs, back to the aorta. Deprived of its blood supply, the rest of the animal dies, so that the heart is functionally denervated and the heart rate varies little if at all. As the heart is a pressure generator, the impedance ("peripheral resistance") of the tubings controls the stroke volume. The level of the reservoir controls the amount of blood returning to the right atrium ("venous return"). In these experiments the aortal and venous pressures and the ventricular volume were measured. The results of these well-known and often repeated experiments are as follows. When the impedance is raised, then the aortic pressure raises also, and the stroke volume decreases, so that blood accumulates in the ventricles, and the size of the heart increases. Due to the "Frank-Starling Law", the heart starts to beat more forcefully, and the stroke volume returns to its previous level after a few beats.

If on the other hand the impedance is reduced, then the aortic pressure decreases also, and the stroke volume increases initially, but as the volume of the ventricle decreases, the stroke volume falls again to the previous level.

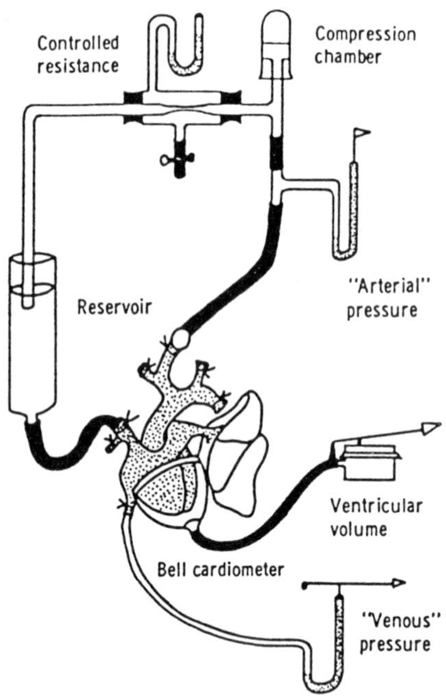

Fig. 1.6. Heart-lung preparation (from Ganong (1973))

Thus, the peripherical resistance controls the pressure in the vascular system, and consequently in the heart. The pressure in the aorta loads the heart during ejection of the blood. In single muscle fibres, this situation is seen in experiments studying *afterload.*

1.5 Myocardial contractility

A third factor exerting a major influence on the stroke volume is myocardial contractility. Sympathethic nerve impulses shift the whole length-tension curve upward and to the left.

Changes in cardiac rate affect myocardial contractility. Also, the intra- and extracellular concentration of Ca^{++}-ions, the Ca^{++}-permeability of the cell membranes, and the action of the β-receptors affect the myocardial contractility. Hypoxia, some drugs and intrinsic depression reduce the contractility.

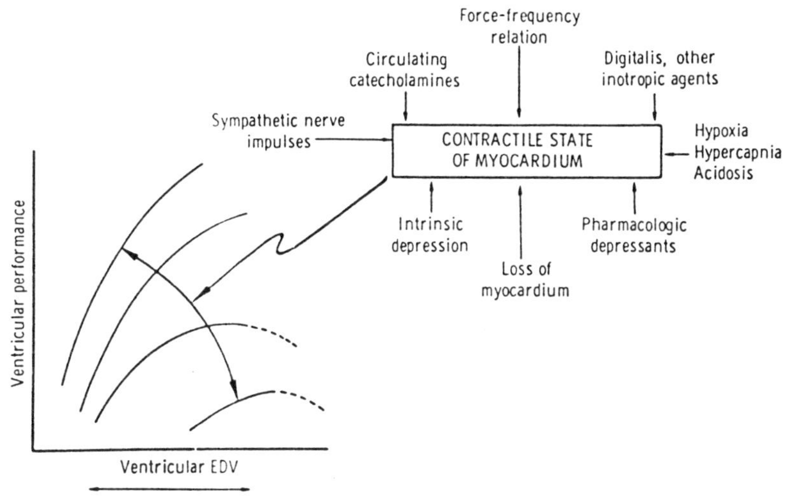

Fig. 1.7. Effects of changes in myocardial contractility on the Frank-Starling curve (from Braunwald 1967 in Ganong (1973))

1.6 Models for the cardiovascular system

The qualitative working of the cardiovascular system as sketched in the previous paragraphs is now reasonably well understood; its quantitative analysis has met with considerable difficulties. One of the major problems is to collect experimental data. As far as the physical characteristics of the system are concerned, three variables have to be experimentally evaluated, namely the "inotropic" variables: pressure, flow and volume. Each of these variables is time dependent, and even with the most careful techniques, the measuring errors *in vivo* will be of the order of five per cent. Other variables, such as tension, cannot be measured directly, and have to be estimated from the three measured variables, so that the errors in these variables are even larger than five per cent. Reproduction of the data is extremely difficult, because of the extrinsic control of the cardiovascular system and the biological distribution.

Nowadays different types of analog models are used to study the circulatory system. The existing models can be divided into hydrodynamic and electrical analogs.

Although hydrodynamic models have features which the electrical analogs do not possess, they are less frequently used because of the difficulties which have to be overcome to construct imitations of the cardiovascular system. For instance, it is difficult to obtain tubes with the same modulus of elasticity, radius and wall thickness as the blood vessels or a pump with the same time-dependent characteristics as the ventricles. It is quite difficult to build into such imitations variables that enable the presetting of changes as in the real world (see (Attinger 1960) and (Janssens 1989)).

The electrical analogs use an idealized description of the cardiovascular system. The arterial and venous branches are divided into only a few (a lumped model) or a large number of segments (an extended model). These segments are represented in terms of ordinary differential equations to permit (analog or digital) computer simulation, which contains only resistors (R), compliances (C) inertances (L) and diodes. The coefficients (R, C, L) of these equations are related to anatomical dimensions of the blood vessels (radius r and length l, modulus of elasticity E, Poisson's ratio σ and wall-thickness h), and characteristics of the blood (μ viscosity and ρ density).

As an example, (Snyder et al. 1968) computed the relationship:

$$R = \frac{81\mu l}{8\pi r^4}, \; L = \frac{9\rho l}{4\pi r^2} \text{ and } C = \frac{3\rho^3(1-\sigma^2)l}{2Eh}. \tag{1.1}$$

A simplified equivalent circuit for a section of an artery or a vein is shown in Figure 1.8. The network shown in Figure 1.8, second figurine, results in a set of approximate linear integral equations for pressure and flow F (*loc. cit.*).

$$P_i = R_p(F_i - F_{i+1})\frac{1}{C}\int [F_i - F_{i+1}]\, dt \tag{1.2}$$

$$F_i = \frac{1}{L}\int (P_{i-1} - P_i - R_s F_i)\, dt\ . \tag{1.3}$$

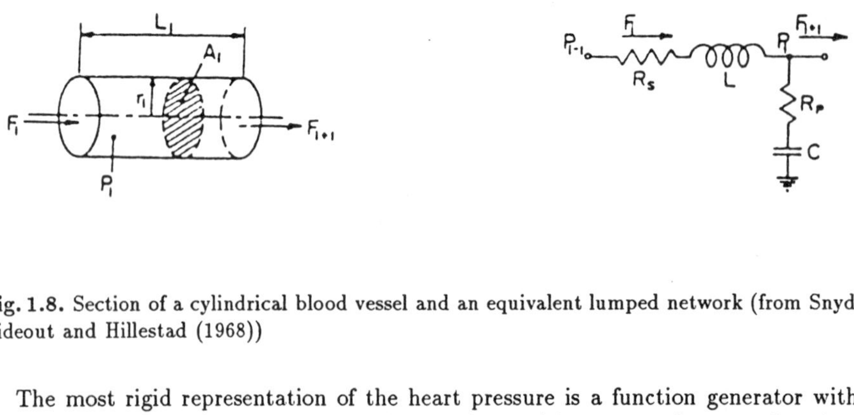

Fig. 1.8. Section of a cylindrical blood vessel and an equivalent lumped network (from Snyder, Rideout and Hillestad (1968))

The most rigid representation of the heart pressure is a function generator with a fixed pressure wave. / This representation can be used in one steady state situation of the cardiovascular system. The representations of the atria and ventricles in *loc. cit.* are appropriate in situations with minor volume changes in the cardio-vascular system, since a time varying compliance is used to provide both ventricular and atrial pumping. The valves are simulated by diodes. The relation between the ventricular pressure, volume and compliance is given by

$$P_{ven} = \frac{V_{ven}}{C_{ven}}. \tag{1.4}$$

In Snyder's simulations a rectified sine-wave was chosen as varying compliance. During diastole the pressure is assumed zero. The beat rate of the heart can be changed

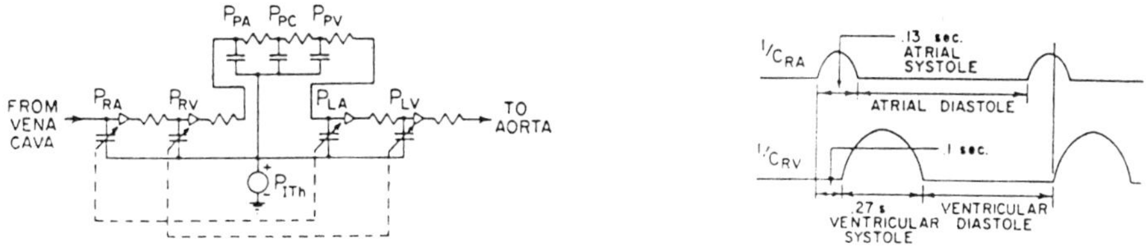

Fig. 1.9. Representation of the heart in (Snyder and Rideout (1969)

by varying the duration of the diastole. The time graph of the compliance is the same for all the systoles. The ventricular volume determines the amplitude of the ventricular pressure during systole. The maximal ventricular pressure or peak systolic pressur (PSP) is proportional to the ventricular volume. This is a linearization of the "Frank-Starling Law", which holds only over small intervals close to the normal working point of the ventricle.

Comments.

1. The changes in contractility are not included in the model by Snyder and Rideout,1969. This is due to the fixing of the wave form of the compliance, and consequently of the wave form of the ventricular pressure during systole. Only the amplitude of the ventricular pressure during systole can be altered by the EDV. The results published in the paper (*loc. cit.*) concerning the response of the pressure and flow waves during tilting of patients should be handled with great care, precisely because the model only includes a linear approximation of the "Frank-Starling Law".

2. Sagawa et.al. improved the heart generator. They determined experimentally the time course of the compliance by dividing the measured ventricular volume by the ventricular pressure, averaged over several beats. This generator is more realistic but the main limitations of Snyder's approach remain. We have used Sagawa's representation up till now, but will define a different one in the following.

Chapter 2 Catastrophes

2.1 Introduction

Catastrophe theory is a mathematical method created by René Thom in 1972. It is particularly applicable where gradually changing forces produce sudden effects. These effects are often called *catastrophes*, because our intuition expects a smooth behaviour. They have nothing to do with catastrophes in the usual sense of the word, but with discontinuous behaviour without a value judgment attached. The ideas of continuity dominated the classical physics from Newton to the General Relativity. However, some physical problems show a discontinuous behaviour such as : phase transition, or buckling of a plane. In biology, medicine and social sciences many kinds of sudden changes caused by smooth alterations in the situation appear, and can be explained with catastrophe theory. The theory depends upon some deep theorems in the geometry of high dimension, which classify the ways that discontinuities can occur in terms of an infinite series of archetypical unfoldings. Thom called these unfoldings catastrophes or catastrophy types. Catastrophes are described using state and control parameters. Thom showed that the catastrophe types with one state parameter x, can be represented as polynomials in x with the control parameters as coefficients. Diffeomorphisms conserve a catastrophe type, even of types with more than one state parameter. However, every type admits a representation in polynomial form, called 'standard unfolding' or 'rational unfolding'. The control parameters control the system, which is thought to be in an equilibrium state. In classical physics, a free friction system tends to a state of (local) minimal energy for given values of control parameters. This state can be an equilibrium state, a limit cycle or a diverging state. For the first two an analogue of these classical interpretations will be given in the following paragraphs. Let E be a polynomial energy function of the order $n+1$ in one state variable x given by:

$$E = \frac{1}{n+1}x^{n+1} - \frac{a_{n-1}}{n-1}x^{n-1} - \ldots - a_1 x - a_0. \tag{2.1}$$

Here,the parameters $a_0, \ldots, a_{n-1}$ are the control parameters. The parameter a_0 can be omitted by shifting the system to the origin of the energy. The critical points of the energy function for a given set of control parameters $(a_1, \ldots, a_{n-1})$ can be found by setting the first derivative of E with respect to x to zero:

$$\frac{d}{dx}E = x^n - a_{n-1}x^{n-2} - \ldots - a_1 = 0. \tag{2.2}$$

For polynomial functions the critical points give the possible local extrema. If the second derivative is positive or negative the energy function has a local maximum or minimum respectively. The local maxima of the energy function are associated with unstable equilibria whereas a local minimum is related to a stable equilibrium. To illustrate the characteristics of the catastrophe theory we examine the case $n = 3$, which is called the cusp catastrophe (type).

2.2 Cusp catastrophe

The energy function is given by:

$$E_{a_2,a_1} = \frac{1}{4}x^4 - \frac{a_2}{2}x^2 - a_1 x, \tag{2.3}$$

and the critical points are given by:

$$\frac{d}{dx}E_{a_2,a_1} = x^3 - a_2 x - a_1 = 0, \tag{2.4}$$

for given control parameters a_2 and a_1. Relation (2.4) between the state variable x and the control parameters a_2 and a_1 is cubic. This cubic relation in x must have at least one and at most three real roots. The nature of the roots depends on the cubic discriminant

$$D = -4a_2^3 + 27a_1^2 \tag{2.5}$$

of the cubic equation. It is well known that, if the discriminant is negative ($D < 0$) there are three distinct real roots, and if the discriminant is positive ($D > 0$), there is only one real root and a conjugate pair of complex roots. If however the discriminant $D = 0$ there are three real roots but two coincide if $a_1 \neq 0$ or $a_2 \neq 0$; if $a_1 = a_2 = 0$ then the only (threefold) root is $x = 0$. The discriminant D divides the (a_1, a_2)-plane into five parts, separated by the so-called bifurcation curve, namely the origin, the shaded part R_3, the part R_1 and the curves B_1 and B_2 (cf. Figures 2.1, 2.2 and 2.3).

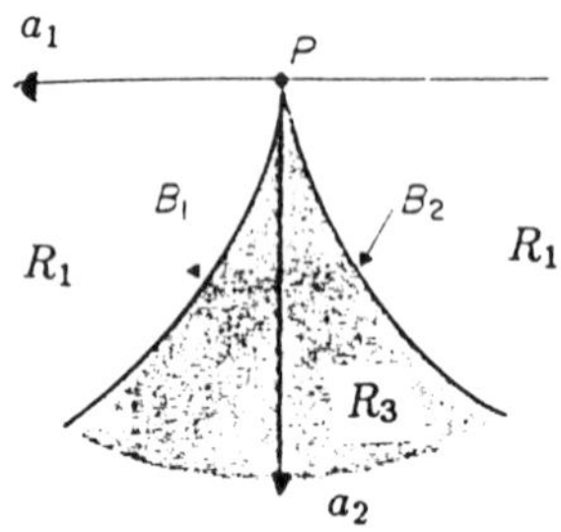

Fig. 2.1. The sign of D as a function of the control parameters(from Poston and Stewart (1978)) a_2 and a_1, in the shaded part $D < 0$ and on the full curve $D = 0$.

In the region R_3 the discriminant is negative, so that there are three real roots. In R_1 where the discriminant is positive, there is precisely one real root. On the bifurcation lines B_1 and B_2 and not in the origin, there are three roots two of which coincide; on B_1 the coincidence occurs for the smaller root, and on B_2 for the larger root. At the origin the three roots are zero.

The equilibrium surface M in (a_1, a_2, x)-space defined by equation 2.4. is called a catastrophe manifold. It has the shape of a folded surface and is illustrated in Figure

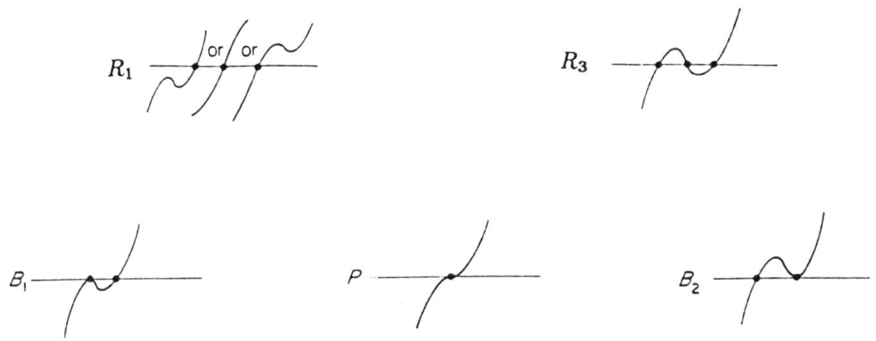

Fig. 2.2. The derivative of E with respect to x is plotted as a function of x (from Poston and Stewart (1978))

2.4. It is called a cusp catastrophe because the bifurcation set B_1 and B_2 has the shape of a cusp.

Let us analyze the quasistatic behaviour of the catastrophe. Fix the value of the control parameter $a_2 = C < 0$. The intersection of the unfolding surface (eq. 2.4) with the plane $a_2 = C$ is a steadily decreasing function of a_1 (see Fig. 2.4). For every a_1 the point (a_1, a_2) lies in R_1, since $D > 0$ and above that point there is only one value of x on the surface.

If the fixed value of the control parameter $a_2 = C > 0$, then the intersection of the surface and the plane is a sigmoid function of a_1. In the region where the discriminant is negative (the shaded part in Fig. 2.1) there are three values for x on the surface, for every a_1. These values correspond to two stable and one unstable equilibrium. The unstable equilibrium lies between the two stable equilibria. The set of all stable equilibria form the upper and the lower sheet of the catastrophe manifold, and the unstable equilibria are the points on the the intermediate sheet (see Figures 2.4 and 2.5).

Normally the system, i.e. the cursor, will be in one of the two stable states; i.e. it moves on one of the two stable sheets. Let a_1 be smaller than

$$-\sqrt{\frac{4}{27} \mid a_2^3 \mid}$$

then $D > 0$ and (a_1, a_2) lies in R_1 if a_2 is kept constant and if a_1 is increased continuously, the cursor (a_1, a_2, x) moves to the left on lower sheet of the surface. If the discriminant D becomes negative, (a_1, a_2) lies in region R_3 and two stable equilibria may obtain. If the increase of a_1 is slow the value of x will change continuously and the cursor will not jump to the other stable sheet. However, when (a_1, a_2) reaches the bifurcation line and if a_1 is still further increased, the cursor (a_1, a_2, x) will jump to the other stable sheet

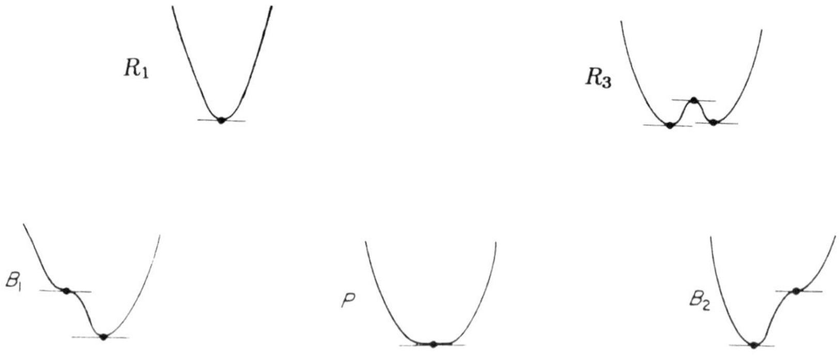

Fig. 2.3. The energy E as a function of x. (from Poston and Stewart (1978))

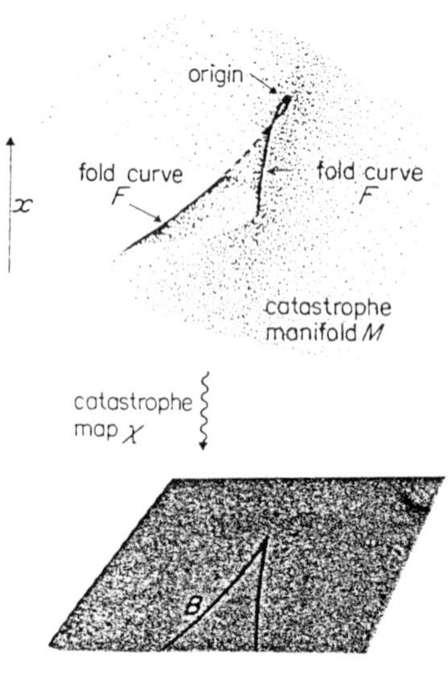

Fig. 2.4. The Cusp catastrophe manifold and the bifurcation set (from Poston and Stewart (1978))

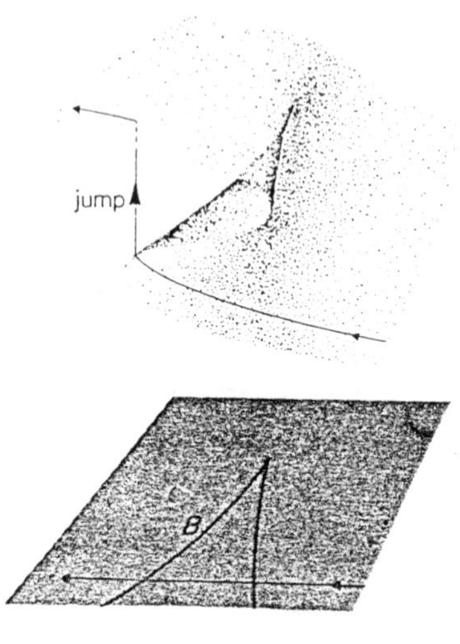

Fig. 2.5. Catastrophic jump (from Poston and Stewart (1978))

of the surface, because only one stable situation is possible in R_1. The jump from one sheet to the other is supposedly fast, so that the time it takes may be ignored. Such a jump is called by Thom and Zeeman "a catastrophic jump".

If the cursor (a_1, a_2, x) lies on the upper sheet and the value of a_1 is decreased continuously, the jump from the upper to the lower sheet will occur when (a_1, a_2) lies on the bifurcation line B_2. Notice that the upward and downward jumps do not occur for the same coordinates of (a_1, a_2). This phenomenon is called hysteresis (by analogy with a similar effect in magnetism).

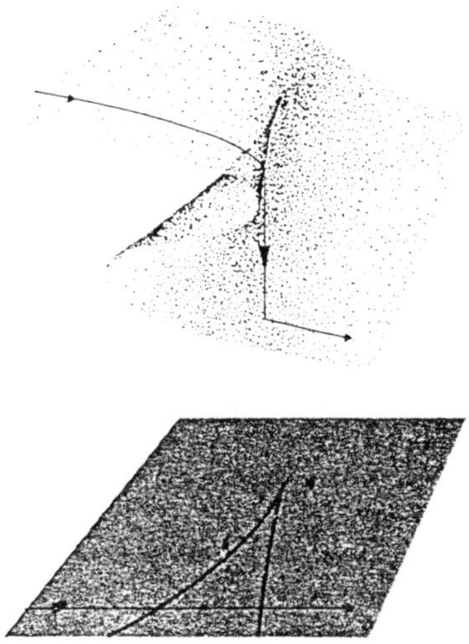

Fig. 2.6. Hysteresis phenomenon (from Poston and Stewart (1978))

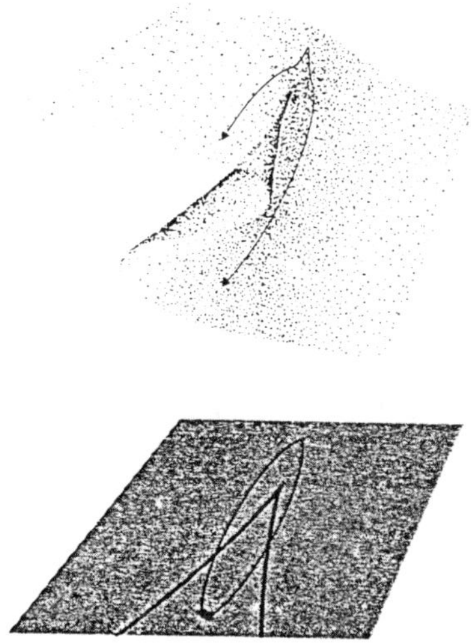

Fig. 2.7. Divergence phenomenon (from Poston and Stewart (1978))

A third phenomenon is bifurcation (also called divergence): slight differences in control paths may (without any jumps occurring) produce large differences in state, even when the former start and end at the same control points.

Finally, the shape of the bifurcation curve in the (a_1, a_2)-plane is a cusp, from which the catastrophe derives its name. This curve has a parametric representation of the form

$$\begin{aligned} a_1 &= -2t^3, \\ a_2 &= 3t^2 \end{aligned} \tag{2.6}$$

where t is a real variable. The representation (2.6) is easily obtained. First the derivative of eq. 2.4 is set to zero, $3x^2 - a_2 = 0$. From this equation and eq. 2.4, a_1 and a_2 are eliminated. It is a so-called (3,2)-cusp, implying that there are higher-order cusps. Indeed, if in (2.1) and (2.2) the positive integer n is any odd number, and if, in addition, all coefficients of (2.1) except a_1 and a_2 are set to zero, then the same elimination leads to a so-called (n,n-1)-cusp in the (a_1, a_2)-plane , obeying the equation

$$\begin{aligned} a_1 &= -(n-1)t^n, \\ a_2 &= nt^{n-1}. \end{aligned} \tag{2.7}$$

It is to be noted that for even values of n, formula (2.7) defines a plane figure that is no cusp (cf. Kuyk and Smits (1990).

Remark. It makes sense to discuss the bifurcation space of the catastrophe manifold defined by (2.2) with its derivative set to zero. It is a $(n-1)$–dimensional hypersurface in $(a_1, ..., a_{n-1}, x)$-space. The parameters a_1 and a_2 can be explicitly expressed in terms of a_3, ...,a_{n-1} and x as follows

$$\begin{aligned} a_1 &= -(n-1)t^n - (n-3)a_{n-1}t^{n-2} - \dots \\ &\quad - 2a_4t^3 - a_3t^2, \\ a_2 &= n\,t^{n-1} + (n-2)a_{n-1}t^{n-3} + \dots \\ &\quad + 3a_4t^2 + 2a_3t. \end{aligned} \tag{2.8}$$

If in (2.8) t is allowed to be a real parameter, then these equations define, for every value of the real vector $(a_3, \dots, a_{n-1})$ a plane bifurcation curve of the catastrophe manifold. These curves are /it isotangent in this sense, that at any point x of the curve, the absolute value of the slope at x is equal to the absolute value of x itself. These curves are known to have only cusp points for singularities, and they reveal a lot of the structure of that part of the manifold that 'lies above' them (see Kuyk and Smits (1989 and 1990)).

Chapter 3 Oscillators

3.1 The harmonic oscillator

The harmonic oscillator is the best known oscillator. Its differential equation is given by the equation:

$$\ddot{x} + x = 0. \tag{3.1.1}$$

This oscillator has no damping at all and a solution for it can be written as:

$$x = A\cos(t + \phi). \tag{3.1.2}$$

The first derivative of x with respect to time is given by:

$$\dot{x} = -A\sin(t + \phi). \tag{3.1.3}$$

with period 2π, phase angle ϕ and amplitude A.

The phase-plane is a useful concept for visualizing the behaviour of a dynamical system. It is an abstract space whose coordinates are the degrees of freedom of the system's motion. The motion of the harmonic oscillator is completely determined by its position x and velocity $\dot{x}$. Its state is thus a point in the phase-plane whose coordinates are x and $\dot{x}$. As the position and velocity oscillate, they follow a path, through the plane. For an harmonic oscillator, the orbit is a closed circle as shown in Figure 3.1.

The oscillator is unstable because an arbitrary small damping changes the type of the orbit. For instance, the introduction of a damping term $\epsilon\dot{x}$ for $\epsilon > 0$, makes the differential equation to read:

$$\ddot{x} + \epsilon\dot{x} + x = 0 \qquad \epsilon > 0. \tag{3.1.4}$$

This equation has no periodic solution if ϵ differs from zero. If on the other hand ϵ is smaller than 2, the solution can be written as an exponential function times a cosine function

$$x = A\cos(\omega t + \phi)\exp(-t/\tau), \tag{3.1.5}$$

where the pseudo-pulsation ω is equal to $\frac{1}{2}\sqrt{4-\epsilon^2}$ and the relaxation time is equal to $\tau = 2/\epsilon$. In the phase plane the trajectory is a spiral which converges to the origin when the time is large in comparison to τ. If the damping term $\epsilon\dot{x}$ is replaced by $-\epsilon\dot{x}$ then for $0 < \epsilon < 2$ the amplitudes of x and $\dot{x}$ are $A\exp(-t/\tau)$ and $A\omega\exp(-t/\tau)$, respectively. They increase with time and the spiral will diverge, because the additional force is in the direction of $v = \dot{x}$.

When yet another term $\epsilon x^2\dot{x}$ is added, we obtain the van der Pol equation.

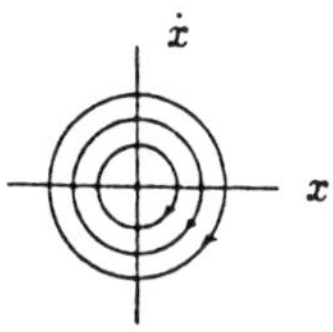

Fig. 3.1. The phase curve of an harmonic oscillator, (from Zeeman (1977))

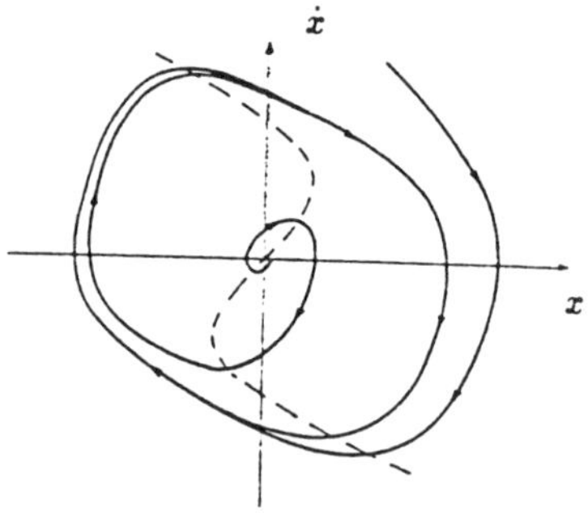

Fig. 3.2. Phase curve of a van der Pol oscillator with small damping (from Stoker (1950))

3.2 Van der Pol oscillator

The differential equation of a van der Pol oscillator is given by:

$$\ddot{x} + \epsilon(x^2 - 1)\dot{x} + x = 0 \qquad \epsilon \text{ small constant} > 0. \tag{3.2.1}$$

This is a stable perturbation of the harmonic oscillator. The damping force is no longer linear, and will be such as to tend to increase amplitude for $|x| < 1$, but decrease it for $|x| > 1$. This differential equation (3.2.1) has the special property that it has one limit cycle. Figure 3.2 indicates what happens.

The spiral near the origin as well as those far away from the origin tend to a single closed curve in the phase space, which corresponds to a periodic solution of equation (3.2.1). / In other words, every solution of eq. (3.2.1) tends, as t tends to infinity, to a periodic solution. It is important that the sign of the damping term be positive for large x, giving a quasi-periodic solution. If the sign of the damping term for large x is reversed, then the limit cycle becomes unstable, and the spiral tends to zero when the initial values of x and $\dot{x}$ lay inside the limit cycle. If and when the initial values do not lay in the limit cycle, then the spiral diverges.

3.3 Van der Pol oscillator with large damping

Consider the van der Pol oscillator but change the small constant ϵ into a large constant K. The differential equation then reads:

$$\ddot{x} + K(3x^2 - a_2)\dot{x} + x = 0 \quad K \text{ large constant} > 0. \tag{3.3.1}$$

Here, the factor 3 is put in for convenience and a_2 is a parameter. Put $\dot{a}_2 = 0$, $x(0) = x_0$ and $\dot{x}(0) = \dot{x}_0$, and let

$$a_1(t) = a_{1,0} - \frac{1}{K}\int_0^t x(t')\, dt', \tag{3.3.2}$$

with

$$a_{1,0} = x_0^3 - a_2 x_0 - \frac{1}{K}\dot{x}_0.$$

The first derivative with respect to t of a_1 is equal to

$$\dot{a}_1 = \frac{-1}{K}x. \tag{3.3.3}$$

equation (3.3.3) is called the *slow equation*. Substituting this into (3.3.2) one gets

$$\ddot{x} + K((3x^2 - a_2)\dot{x} - \dot{a}_1) = 0. \tag{3.3.4}$$

Integrating this differential equation we get

$$\dot{x} + K(x^3 - a_2 x - a_1) = 0. \tag{3.3.5}$$

The integration constant is zero by the choice of the parameter $a_{1,0}$. Equation (3.3.5) is called the *fast equation*. On the manifold the equation $x^3 - a_2 x - a_1 = 0$ holds and hence $\dot{x} = 0$. But due to equation (3.3.3) the value of a_1 will change and so will the value of x. This means that the cursor (a_1, a_2, x) will be in the neighbourhood of the slow manifold. (Near means to the order of $1/K$.) Figures 3.3 and 3.4 illustrate the cases. If the value of x is positive the value of a_1 will decrease, until (a_1, a_2) lies on the bifurcation line B of the slow manifold. The value of a_1 will decrease further because of eq. (3.3.3) and $\dot{x}$ will have very large negative value, so that the value of x will decrease rapidly. If the value of x is negative then the value of a_1 will increase, and the cursor will be attracted by the other sheet of the manifold. The steepness of the trajectory in the (a_1, x)-plane depends on the value of K. / If K tends to infinity a "relaxation oscillation" is obtained. In this case, the jump from one sheet of the manifold to the other will be almost instantaneous, and its graph in the (a_1, x)-plane is a straight line parallel to the x-axis. The other part of the orbit will be very close (of the order of $1/K$) to the slow manifold. Whence the name *attractors* for the two slow manifolds.

The name "relaxation oscillation" was given by van der Pol presumably because the oscillator (3.3.1) exhibits two distinct and characteristic phases, viz. one during which energy is stored up slowly and another one during which energy is discharged nearly instantaneously after a certain critical threshold has been attained.

The period T is obviously given by the line integral

$$T = \oint \frac{da_1}{\dot{a}_1} = \oint \frac{da_1 dt}{da_1}, \tag{3.3.6}$$

and, substituting eq. (3.3.3), by the formula

$$T = K \oint \frac{da_1}{x}, \tag{3.3.7}$$

where $\oint$ denotes the line integral of the cycle. Neglecting the time to jump from one sheet to the other the period can be approximated by:

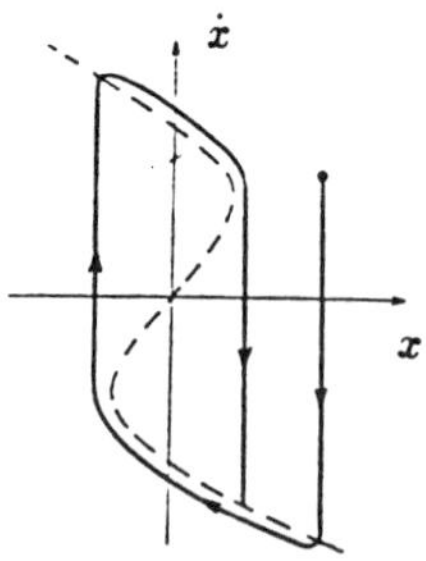

Fig. 3.3. Phase curve of a van der Pol oscillator with large damping (from Stoker (1950))

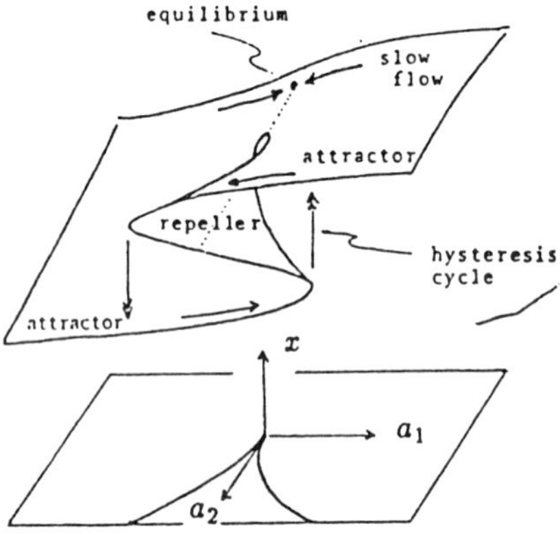

Fig. 3.4. Van der Pol with large damping is a cusp catastrophe (from Zeeman (1977))

$$T \approx 2K \int_{x_1}^{x_2} \frac{d(x^3 - xa_2)}{x} = 2K(-\ln(x)a_2 + \frac{3}{2}x^2)\Big|_{x_1}^{x_2}$$
$$= 2K(-\ln(2) + \frac{3}{2})a_2, \tag{3.3.8}$$

where

$$x_1 = \sqrt{\frac{a_2}{3}} \quad \text{and} \quad x_2 = 2x_1,$$

so that $T \approx 1.614Ka_2$.

In Figure 3.5 the solution of the van der Pol oscillator is plotted as a function of time for different values of ϵ (or K). When ϵ is small the closed solution curve in the phase plane which results will be nearly a circle and the corresponding motion will be nearly a simple harmonic motion of a definite amplitude, as in Figure 3.1. As ϵ (or K) is increased, the limit state that is approached for large t deviates more and more from a simple harmonic motion. The initial values for x and a_1 are zero. The distortion of the sine wave increases markedly with increasing ϵ (or K). The period increases and the shape of the oscillations becomes jerkier when the value of ϵ (or K) increases.

If a_2 is allowed to vary with time then the oscillator (3.3.5) lets the cursor move on (or close to) the canonical cusp (catastrophe) surface. The slow manifolds are the sections of this surface given by the planes $a_2 = constant$. The fast equation acts as a catastrophe dynamic for the variable x, making the upper and lower sheets (where $3x^2 > a_2$) into

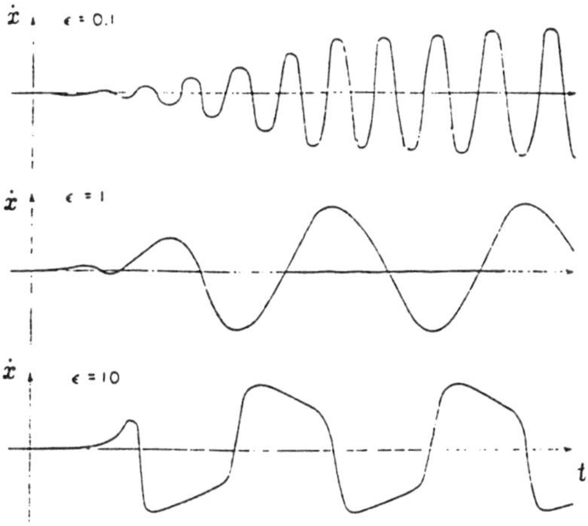

Fig. 3.5. Changes in character of the oscillations with increase in nonlinearity (from Stoker (1950))

attractors, and the middle sheet (where $3x^2 < a_2$) into a repeller. he fast dynamics do three things:

Firstly, it rapidly carries any point towards (or close to) the attracting surface;

secondly, it holds the point near the surface for as long as possible; and

thirdly, when this is no longer possible, then the dynamic parameter causes a catastrophic jump to the other attracting sheet.

The slow dynamics moves a_1 to the left if x is positive and to the right if x is negative. The only external parameter which can be set is a_2. This parameter is often called the *splitting parameter.*

If this parameter is allowed to change during the process, and if the slow equation may remain the same then the second order equation (3.3.1) is no longer equivalent with the two differential equations (3.3.5) and (3.3.3), even though they still describe an oscillating system. If a_2 is a linear function of x then the solution of the differential equations remains periodical. But if a_2 is an explicit periodic function of time t, the differential equation (3.3.5) is no longer free. The solutions are in general no longer periodic, even ' chaos' may occur. Using these ideas, other types of trajectories are possible, e.g. put $a_2 = 2 + x$ in (3.3.5) to get

$$\dot{x} + K(x^3 - (2 + x)x - a_1) = 0. \tag{3.3.9}$$

The simultaneous solution of eqs (3.3.3) and (3.3.9) as plotted in Figure 3.6 shows a periodic behaviour. However, the period can be calculated using equation (3.3.6), but can not be approximated by the formula $T \approx 1.614Ka_2$.

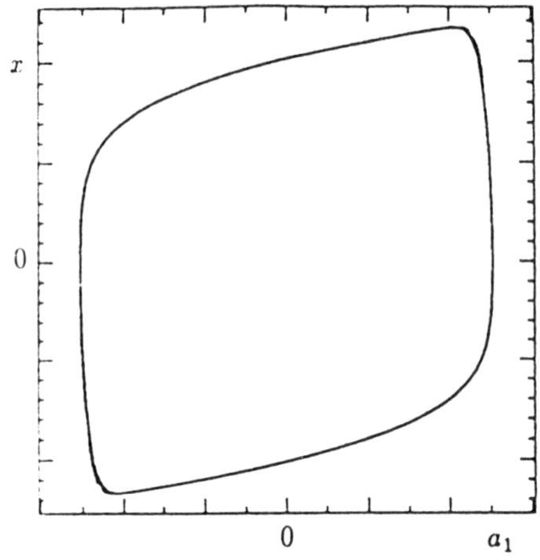

Fig. 3.6. Simulation of an oscillating system with slow equation (3.3.9)

Thus, if a_2 is an explicit function of time t, e.g. $a_2 = \sin(t)$, so eq. (3.3.5) becomes

$$\dot{x} + K(x^3 - \sin(t)x - a_1) = 0, \tag{3.3.10}$$

then the phase curve plotted in Figure 3.7 shows chaotic behaviour and the trajectory does not tend to a simple limit curve as t becomes large.

In order to decrease the period of the system we introduce a time scaling parameter S into equations (3.3.3) and (3.3.5), to obtain

$$\dot{a}_1 = \frac{-S}{K}x \tag{3.3.11}$$

and

$$\dot{x} = SK(x^3 - a_2x - a_1). \tag{3.3.12}$$

Under this scaling parameter the shape of the (a_1, x)-graph does not essentially modify the time graphs; they only show a scaling in the t-direction. The equations (3.3.11-12) are still equivalent to equation (3.3.1) where t is replaced by St.

3.4 Generalization of the degree of the oscillator

We have already remarked that the degree of the cusp in the control plane can be set to any odd number n (cf. formula (2.7)). This implies that one can 'generalize' the present van der Pol oscillator to any odd degree n simply by replacing x^3 with x^n in (3.3.5), and accordingly, replacing $3x^2$ with nx^{n-1} in (3.3.4). Then, with a slight and obvious change, all the formulas, even (3.3.9) and (3.3.10) of the previous paragraph, remain valid for the generalized oscillator.

It has already been remarked that for even n the formulas (2.7) define an (n-1)-dimensional surface whose bifurcation curve is not a cusp figure around the origin. However, a surface which has, at least around the origin, the shape of a cusp, seems to be the right geometrical object for generalizing the van der Pol oscillator. This is one reason why the differential equation

$$\ddot{x} + K(2nx^{2n-1} - \ldots)\dot{x} + x = 0. \tag{3.4.1}$$

obtained by substituting $2nx^{2n-1}$ for $3x^2$ in (3.3.4) does not seem to work if one wants to find a higher order catastrophe model for the ventricular pump in what follows. This consideration is supported by the following analytic argument. Now, since the leading degree of the polynomial in (3.4.1) is now odd, we see that for large negative values of x, the term of $\dot{x}$ is not damping at all. Thus, if x happens to reach a sufficiently large negative value, the system diverges, loosing its periodicity. In Chapter 7 we study the butterfly catastrophe which is based on a polynomial of the fifth degree.

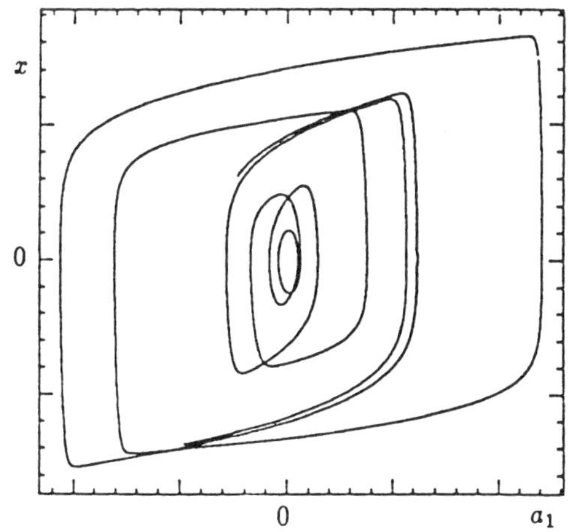

Fig. 3.7. Chaotic behavior of the oscillator with slow equation (3.3.10)

Chapter 4 Records of the Mammalian Ventricular Cycle

4.1 Representations of the ventricular events

As indicated already, pressure-time graphs and ventricular performance- EDV plots are customarily used to study ventricular function. The first include the time as a parameter whereas the second are snapshots. But other representations have also been used. (A. Bove and P. Lynch 1970) determined the force-velocity-length relationship in the ventricle of a dog. Here the time is not an explicit but an implicit variable. Figure 4.1 shows a typical three-dimensional diagram from a mongrel dog (LVP 110 mm Hg, rate 65 puls/min). In this diagram the time is eliminated but the suggestion of time is preserved by plotting points representing time intervals of the same length; the time difference between two dots or tick marks is 14.8 ms. Incidentally, most of our own diagrams are made up of such tick marks only.

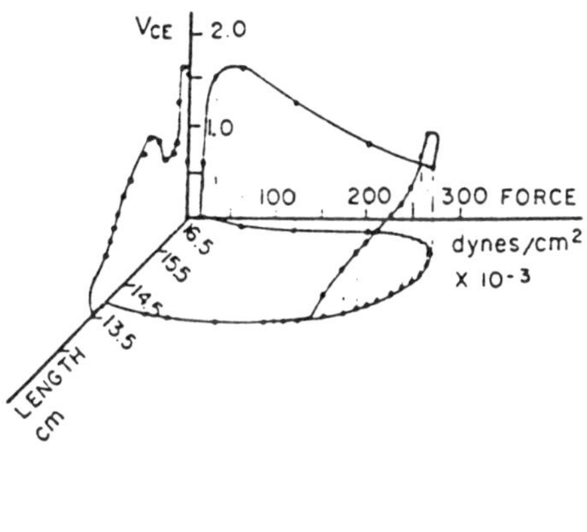

Fig. 4.1. A typical three-dimensional diagram taken from a normotensive dog (from Bove and Lynch (1970))

The present publication, as well as all the published investigations referred to, by necessity, employ graphs in two or three dimensions, even if a full understanding of the ventricular events warrants representations in higher dimension. In order to get a full graphic picture of the ventricular processes, we have derived, by composition and deduction from recorded graphs in the literature, graphs other than the customary ones.

The time graphs of the ventricular pressure and ventricular volume, plotted in textbooks, represent mean values over the human population. We obtained the time flow graph by numerical differentiation of the time ventricular volume of Figure 1.3 (from Ganong (1973)). We eliminated the time between these graphs to obtain other graphical pictures which also show characteristics of the cardiac cycle. Doing so, we obtained volume-pressure plot (Fig. 4.2), volume-flow plot (Fig. 4.3) and flow-pressure plot (Fig. 4.4).

Thus, from a digitized graph of Figure 1.3, (borrowed from (Ganong 1973)) , we derived a two-dimensional volume-pressure plot of the left ventricular cycle, see Figure 4.2. The time between two consecutive tick marks is 8 ms. In steady state, the diagram is a closed curve, showing a steep jump and a steep fall of the pressure at points where the volume looks almost constant. These events are known as the *isometric phases.* The behaviour of the pressure over a full period is as follows. After highest pressure the volume decreases (ejection phase), while it increases during the low pressure phase. The high pressure phase forms part of the ventricular systole. The low pressure phase is part of the diastole. The system makes a counterclockwise cycle in the volume graph.

When the ventricular volume is maximal (EDV or end diastolic volume; the ventricle contracts, while the valves are closed) the pressure jumps steeply up to a level where it overtakes the aortic pressure, so that the aortic valve opens. Then, even as the pressure is still rising, the volume decreases rapidly (systole, large flow) until the pressure slowly decreases to take the flow down to zero up till the point where the ventricular volume reaches its minimal value (ESV or end systolic volume). Then the aortic valve closes and the pressure falls rapidly. During the low pressure phase two different patterns can be observed. The first pattern is a rather fast filling which slows down when the volume is about 110 ml. Then a bulge appears, which is due to the action of the atrium, and next, after the bulge, the ventricular pressure oscillates during the period that the atrium is emptied into the ventricle. The tick marks on the curve represent time intervals of 8 ms. As one can see, the tick marks are sparse when the pressure jumps or falls rapidly. They are close to each other at the end of the systole, and just before and during atrial action. The diagram suggest pressure jumps from a low to a high level and vice versa at constant volume.

The volume-pressure plot gives only a partial insight into the activity of the ventricle. Using numerical differentiation (using the PASSAGE program), we have been calculating from Figure 1.3 the flow through the ventricle. We take the flow to be positive when the volume in the ventricle increases.

In Figure 4.3 the volume-flow diagram is plotted. This gives yet another view of the cyclic activity of the ventricle. As for the volume-flow plot, Figure 4.3, the parts above and below zero show a completely different behaviour. The system makes a clockwise cycle in the volume-flow plot. From EDV the flow jumps to a large negative value. When the aortic valve opens (due to the sudden pressure rise in the ventricle) the blood is ejected into the aorta. Then the flow returns to zero, and the volume decreases. When ESV is reached the pressure falls to a low level and the atrio-ventricular valve (AV) opens, and the flow increases as well as the volume. After a while the flow decreases again to zero. The next phase is due to the emptying of the atrium into the ventricle, which causes an additional volume increase, and EDV is reached again. The concentration of tick marks, near the ESV and before the atrial activity, shows a point where the system remains unchanged for a while.

We are left to look at the flow-pressure diagram (see Figure 4.4). Once again the two pressure events appear. The cluster of recorded points in the low level, during diastole, is due to the filling of the ventricle and the atrial activity (small loop). During the pressure rise the flow oscillates around zero. It is a point of discussion as to what happens to the flow during this isometric phase. Some investigators believe that the volume does

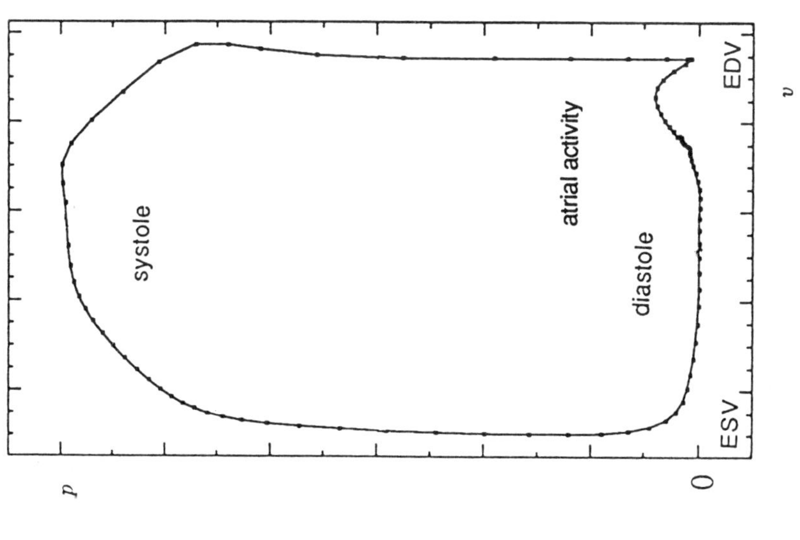

Fig. 4.2. Volume-pressure plot of the ventricle

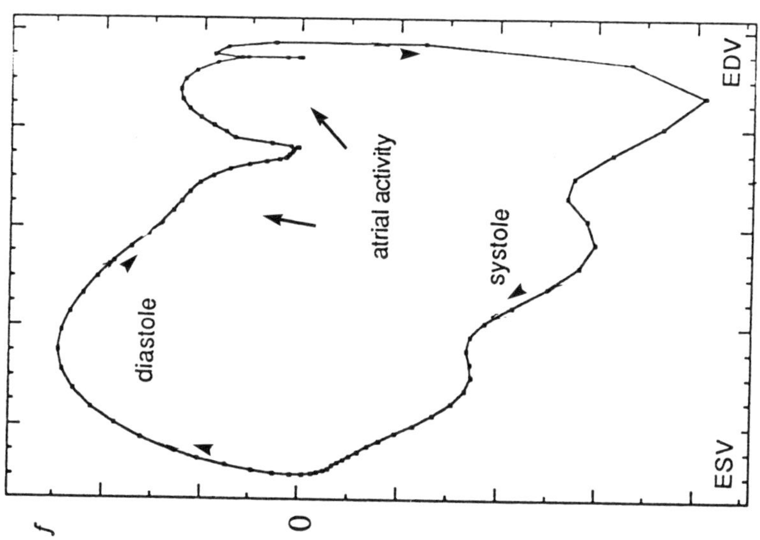

Fig. 4.3. Volume-flow plot of the ventricle

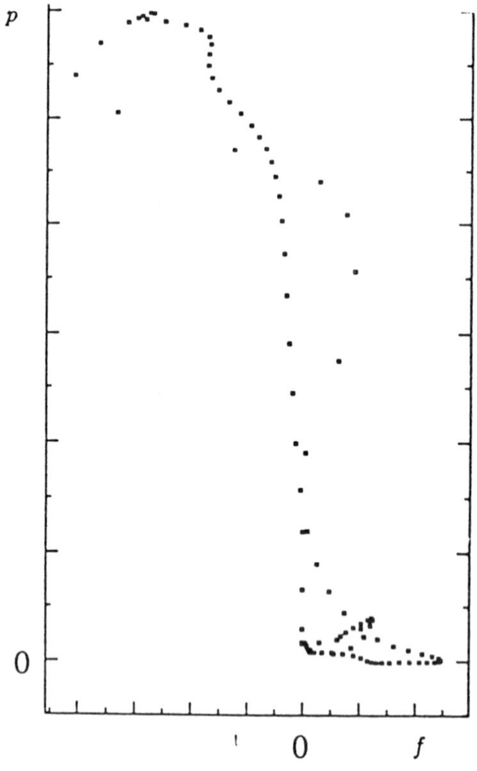

Fig. 4.4. Flow-pressure plot of the ventricle

not change during this rise, and explain the oscillation as an artefact due to measuring errors, while others insist on a change of ventricular volume during contraction due to the curving of the (tricuspidal) AV-valves. The vascular part of our simulation is made up in such a way that during this phase, the volume is constant. However, things can be made to happen differently if so desired. The high pressure level is a part of the ventricular systole. When the heart muscle relaxes, the aortic valve closes and, over the period that the flow is zero, the pressure falls rapidly to the low level. Finally, a more global view is shown in the three-dimensional plot of Figure 4.5, where the pressure is plotted against volume and flow.

The volume-flow-pressure plot (Figure 4.5) shows a jerky orbit in the volume-flow-pressure space, which can be simplified when the action of the atrium is eliminated.

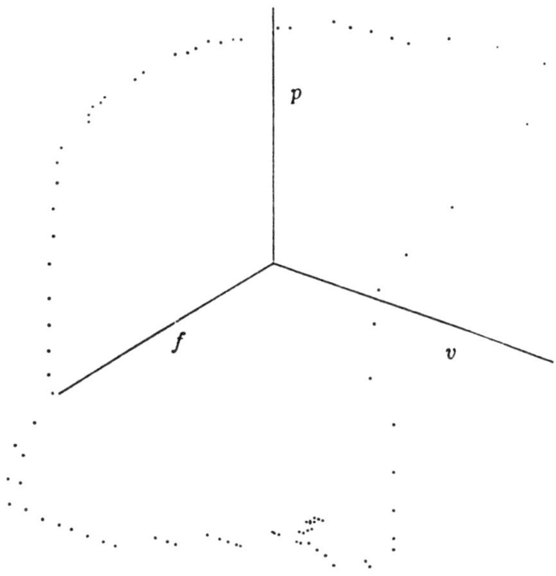

Fig. 4.5. Volume-flow-pressure plot of the ventricle

In the following chapters we only consider the action of the ventricle and neglect the contribution of the atrium.

4.2 Atrial activity

The heart has two sides, left and right. In each side there is an atrium and a ventricle. Each atrium has three functions. The major function is collecting the blood from the veins during ventricular systole. The second function is to transport the blood through the AV-valves into the ventricle. And last but certainly not least the atrium incites and stimulates the contraction of the heart. The sino-atrial-node (SA-node) is located in the right atrial wall. As explained earlier, the ventricle remains beating even when the SA-node is not functioning. Other heart cells (e.g cells located in the AV node or on the bundle of His) may take over the stimulations of the heart muscle. The pumping action of the atria is not essential, although it controls the filling of the ventricle which in turn controls the cardiac output. Thus the atrial action may tune the performance of the heart. Even without atrial contraction the heart remains beating. A frog heart for example does not have atria and has only ventricles. From the point of view of evolution, the atria developed rather late. In fact, the growth of the atria into the heart can be seen as putting the heart onto a higher level of development, since it improves its regulation abilities. But as the system has thus grown more complex, the timing between atria and ventricle becomes a new issue that may lead the heart astray. When both atrium and ventricle interact in the right way the cardiac performance increases. On the other hand, a dephased atrial contraction may obstruct the blood flow from the veins to the ventricles, which may cause a spectrum of new diseases such as arrhythmias and heart blocks of sorts.

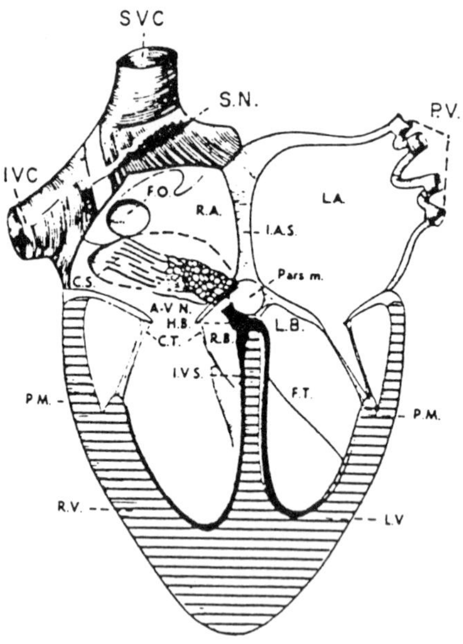

Fig. 4.6. Scheme of the heart. S.N., sino-atrial node; P.V., pulmonary veins; R.A., L.A., right and left atria; S.V.C, I.V.C superior and inferior vena cava (from Schütz in Brobeck (1979))

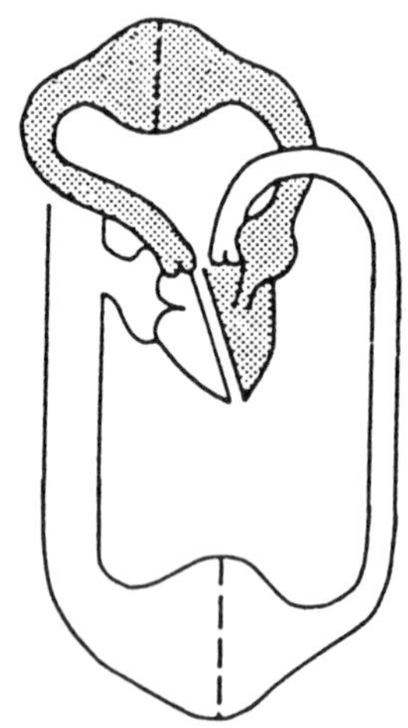

Fig. 4.7. Scheme of systemic and pulmonary (shaded area) system (from Milnor in Brobeck (1979))

The atrium will be omitted in our simulation model, by diverting the atrial functions to other quarters of the cardio-vascular system. Thus, the collecting and transporting functions of the atrium will be placed in the vascular part of the model, and the stimulation function is attached immediately to the ventricle by including it in the catastrophe machine. This simplification gives a better understanding of the ventricular function. Note that we have altogether omitted the endocrine function of the walls of the atrium, but this function, as well as the other three can be explicitly taken up in the model at a later stage if desired.

4.3 The pulmonary circulation

From the right ventricle, blood is pumped into the lungs. In the alveoli of the lungs oxygen binds to the hemoglobin of the blood cells, dissociating carbon dioxide and hemoglobin. Then the oxigen-rich blood is collected in the pulmonary veins where it is stored before flowing to the left heart. The major function of this circulation is to refresh the oxygen of the blood and to eliminate the carbon dioxide. The pulmonary system is not fundamentally different from the systemic circulation from a circulation point of view. In this essay we simulate only the systemic circulation; the pulmonary system may be included later if desired.

Chapter 5. Simulation of the Cardiovascular System

5.1 General explanation of the scheme

Since we want to focus attention on the action of the left ventricle joined to the vascular system, we suppress its pulmonary subsystem and delete the atrium for simplicity. The block diagram of the cardiovascular model as we conceive it is shown in Figure 5.1. The general scheme of the model simulating the cardiac ventricle is as follows.

Five blocks build the vascular system, while one block represents the cardiac ventricle. Starting from the ventricle, the blood flows through the transfer-block, the aortal system, the arterial system, and via the periphery to the venous system and then back again into the ventricle. The transfer-block represents the aortic valve. The other names speak for themselves. The dotted box encloses the AV-valve of the left heart.

Since we simulate the vascular part of the block diagram electrically, the five blocks are made up of electric components, such as resistances, inductances, compliances and diodes. The details are shown in Figure 5.2.

The two ventricular valves and the total system of vascular valves are represented by diodes with kernel letter D. The vascular indices of resistances, inductances, compliances, pressure and flow are represented by their electric analogues with capital kernel letters R,L,C,P and minuscule kernel letter f, respectively. The plain subscripts to all these letters are the first three letters of the names of the boxes in Figure 5.1 to which they refer. This vascular model is a simplification of the scheme by Snyder and Rideout (1969) and Sagawa et al. Since we do not aim to simulate the pressure and flow in the whole peripheral system, but rather simulate realistic pressure and flow waves in the aorta, we have reduced the number of blocks used by these authors to only five. Parallel resistances are also omitted in most cases. However, an inductance is introduced in the branch parallel to the aortal block. The modification thus requires that the first parallel branch contains also a resistance R_{aor} and an inductance L_{aor}. The most relevant components are the venous capacitance C_{ven}, the peripherical resistance R_{per} and the valves D_{per}, D_{ven} and D_{tra}. The capacitance C_{ven} controls the available blood for the next beat. The peripherical resistance R_{per} controls the mean flow through the system. The diodal valves prevent back flow and allow pressure differences between the blocks.

In both Rideout's model, and that of Sagawa, a variable compliance mimics the ventricular pump. If the ventricular pressure and volume are denoted by the minuscules p and v, respectively, then Sagawa determined a time dependent compliance as a function of time simply by dividing a recorded volume-time function by the pressure-time function ($C = v/p$). The obtained compliance was periodically extended.

The resulting system includes a linear "Frank-Starling Law", because the pressure is equal to $p = v/C$. In the next chapters we present a catastrophic model of the ventricular pump, but before doing so, we study the electric scheme and test the vascular part of the model using our own compliance generator (see Figure 5.2 and Section 5.4). To be able to include myocardial contractility, the wave form of our compliance is so constructed that it can be changed at will.

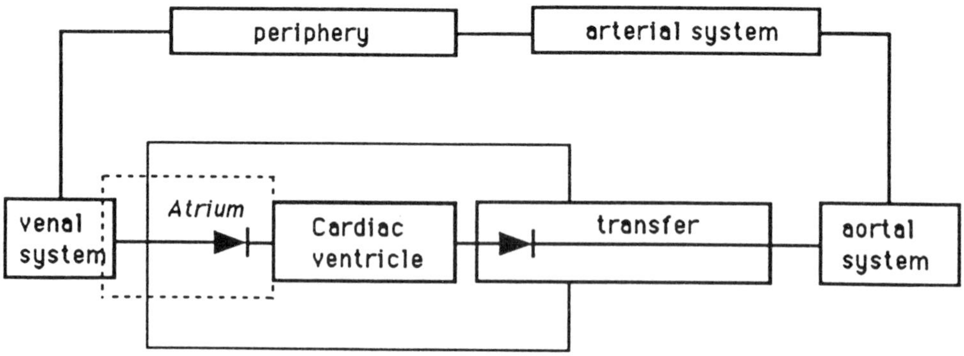

Fig. 5.1. Block diagram of the cardio-vascular system

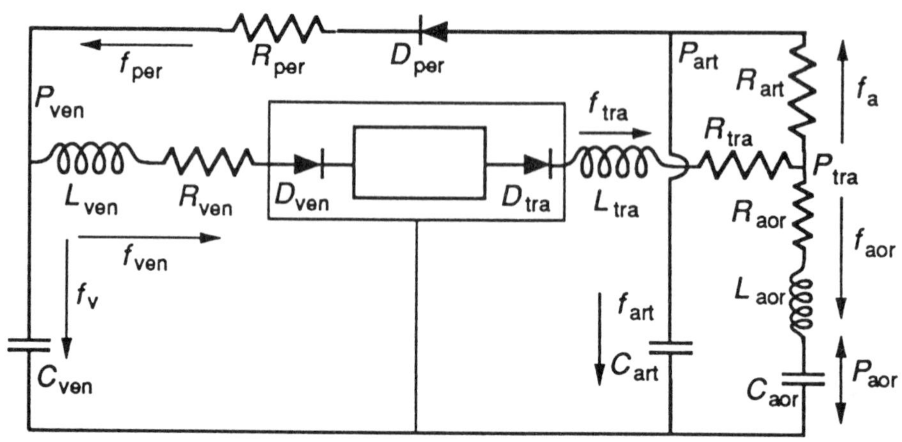

Fig. 5.2. Electrical scheme of the vascular model

5.2 Mathematical formulas for the vascular model

The ventricular pressure p opens the outlet (or aortic) valve as soon as it is big enough to open the diode D_{tra} (cf. Figure 5.2). The flow through the valve is f_{tra}. Due to the valve the flow is always positive or zero. Using Kirchoffs Laws, the pressure p at the entrance of the valve can be expressed in the "tra"nsfer magnitudes by means of the formulas:

$$p = L_{\text{tra}}\frac{df_{\text{tra}}}{dt} + R_{\text{tra}}f_{\text{tra}} + P_{\text{tra}} \qquad \text{if} \qquad f_{\text{tra}} > 0, \tag{5.2.1}$$

and if $f_{\text{tra}} \leq 0$ then P_{tra} and p are independent, since the diode D_{tra} prevents backflow.

When the valve is closed then both p and P_{tra} are independent. The transfer pressure P_{tra} can, in turn, be expressed in terms of the "aor"tal and "art"erial magnitudes. The pressures at the entrance of the aortic and arterial boxes are equal to the pressure P_{tra}, and the sum of the flows f_{aor} and f_{art} is equal to the flow f_{tra}

$$P_{\text{tra}} = R_{\text{aor}}\ f_{\text{aor}} + L_{\text{aor}}\frac{df_{\text{aor}}}{dt} + \frac{1}{C_{\text{aor}}}\int f_{\text{aor}}dt \tag{5.2.2}$$

$$P_{\ rmtra} = R_{\text{art}}f_{\text{a}} + P_{\text{art}} \tag{5.2.3}$$

$$P_{\text{art}} = \frac{1}{C_{\text{art}}}\int f_{\text{art}}dt. \tag{5.2.4}$$

Here, the arterial pressure P_{art} ties in with the diodal nature of the periphery and veins by

$$P_{\text{art}} = \ R_{\text{per}}\ f_{\text{per}}\ + P_{\text{ven}}\ \text{if}\ f_{\text{per}} > 0, \tag{5.2.5}$$

and if $f_{\text{per}} \leq 0$ then P_{art} and P_{ven} are independent, since the diode D_{per} prevents backflow.

The ventricular pressure P_{ven}, in turn, ties in with the diodal nature of the inlet valve D_{ven} by means of the formulas

$$P_{\text{ven}} = L_{\text{ven}}\frac{df_{\text{ven}}}{dt} + R_{\text{ven}}f_{\text{ven}} + p \quad \text{if}\ f_{\text{ven}} > 0, \tag{5.2.6}$$

and if $f_{\text{ven}} \leq 0$ then P_{ven} and p are independent since the diode D_{ven} prevents backflow. The venous pressure is given by

$$P_{\text{ven}} = \frac{1}{C_{\text{ven}}}\int f_{\text{v}}dt. \tag{5.2.7}$$

Finally, for the different flows and for the ventricular volume v we get the formulas

$$f_{\text{per}} = f_{\text{ven}} + f_{\text{v}} \qquad f_{\text{tra}} = f_{\text{aor}} + \ f_{\text{a}} \tag{5.2.8a, b}$$

$$f_{\text{a}} = f_{\text{per}} + f_{\text{art}} \qquad f = f_{\text{ven}} - f_{\text{tra}} \tag{5.2.8c, d}$$

$$v = \int f dt. \tag{5.2.8e}$$

The set of integro-differential equations (5.2.1-8) are non-linear because of the presence of the valves (diodes), and cannot be solved analytically. But the equivalent set of integral equations can be solved numerically, if the function p is given and appropriate values are assigned to the electric components of the scheme. The equivalent set of integral equations is shown in the next paragraph.

5.3 Set of integral equations

From eqs. (5.2.1) and (5.2.3) we obtain

$$f_{\mathrm{tra}} = \max(\frac{1}{L_{\mathrm{tra}}}\int (p - R_{\mathrm{tra}} f_{\mathrm{tra}} - P_{\mathrm{tra}})\, dt, 0) \tag{5.3.1}$$

$$P_{\mathrm{tra}} = R_{\mathrm{art}}\ f_{\mathrm{a}} + P_{\mathrm{art}}. \tag{5.3.2}$$

The right hand side of eq. 5.3.1 means that as long as the integral of the first component of the vector (intergrand) is positive then it is f_{tra}, and it is zero otherwise. Notice that in eq. (5.3.1) the flow f_{tra} through the valve is not an explicit function, it appears also in the integrand. From eqs. (5.2.2) and (5.2.4) we obtain

$$f_{\mathrm{aor}} = \frac{1}{L_{\mathrm{aor}}}\int (P_{\mathrm{aor}} - R_{\mathrm{art}} f_{\mathrm{a}} - R_{\mathrm{aor}} f_{\mathrm{aor}} - P_{\mathrm{art}})\, dt \tag{5.3.3}$$

with

$$P_{\mathrm{aor}} = \frac{1}{C_{\mathrm{aor}}}\int f_{\mathrm{aor}} dt \tag{5.3.4}$$

$$f_{\mathrm{art}} = f_{\mathrm{tra}} - f_{\mathrm{aor}} - f_{\mathrm{per}} \tag{5.3.5}$$

$$P_{\mathrm{art}} = \frac{1}{C_{\mathrm{art}}}\int f_{\mathrm{art}}\, dt. \tag{5.3.6}$$

From eq. (5.2.5) we obtain

$$f_{\mathrm{per}} = \max(\ \frac{1}{R_{\mathrm{per}}}\ (P_{\mathrm{art}} -\ P_{\mathrm{ven}})\ , 0). \tag{5.3.7}$$

From eqs. (5.2.6) and (5.2.7) we obtain

$$f_{\mathrm{ven}} = \max(\frac{1}{L_{\mathrm{ven}}}\int (P_{\mathrm{ven}} - R_{\mathrm{ven}} f_{\mathrm{ven}} - p)\ dt\ , 0) \tag{5.3.8}$$

$$P_{\mathrm{ven}} = \frac{1}{C_{\mathrm{ven}}}\int f_{\mathrm{v}}\, dt. \tag{5.3.9}$$

Equation (5.2.8) remains the same

$$f_{\mathrm{per}} = f_{\mathrm{ven}} + f_{\mathrm{v}} \qquad f_{\mathrm{tra}} = f_{\mathrm{aor}} + f_{\mathrm{a}} \tag{5.3.10a, b}$$

$$f_{\mathrm{a}} = f_{\mathrm{per}} + f_{\mathrm{art}} \qquad f = f_{\mathrm{ven}} - f_{\mathrm{tra}} \tag{5.3.10c, d}$$

$$v = \int f\, dt. \tag{5.3.10e}$$

For our simulation we used values of the lumped electric components of the scheme of Snyder and Rideout (1969). We also scaled these values, for a reason to be explained later.

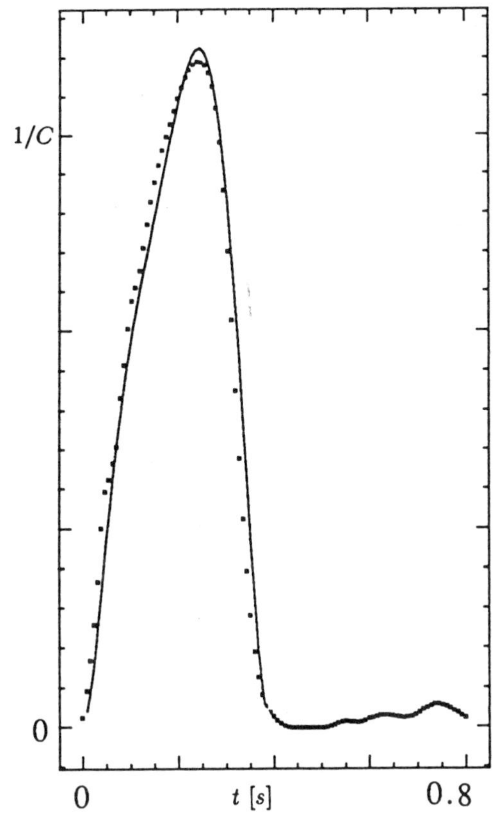

Fig. 5.3. The dotted graph is the reciprocal of the ventricular compliance. The solid curve, covering a density of dots at the right, is the seventh order polynomial approximation of the reciprocal compliance during systole.

5.4 Ventricular compliance generator

As a first test of our cardio-vascular model we used a model after Rideout to simulate the ventricular pressure. The ventricular block in the scheme of Figure 5.2 is first replaced by a new variable compliance. This compliance is based on a digitalization of Figure 1.3. The pressure and volume data of the left ventricle so obtained are then employed to calculate the ventricular compliance $C = v/p$. The graph of the reciprocal of the compliance is shown in Figure 5.3 The reciprocal of the compliance is low during ventricular diastole and atrial systole. During the ventricular systole the reciprocal of the compliance increases, and decreases rapidly after systole. In our model, we keep the compliance as it is at the end of systole through diastole and during atrial systole. This cycle is periodically extended. The period is T =0.80 s, and the time interval between the points is 8 ms. To implement this generator in a simulation programme we need to express it in a functional way. A common way is to use a Fourier sum

$$c_0 + \sum_{i=1} (s_i \sin(2\pi i\omega t) + c_i \cos(2\pi i\omega t)). \tag{5.4.1}$$

Here ω is the pulsation of the cycle and is equal to $2\pi/T$. To calculate the Fourier coefficients s_i and c_i, we fit a polynomial through the systolic and through the reciprocal of the compliance data points. The polynomial coefficients can then be calculated using, for instance, the program PASSAGE. This program fits for any n a polynomial of order n through the data points. In table 5.1tThe standard error is shown as a function of n.

n	1	3	5	7
Stand. err.	0.58	0,13	0.06	0.04

Table 5.1 Standard error as function of the order of the fitting polynomial

Clearly, a fairly good approximation is obtained for n equal to six or seven. The approximation of the reciprocal of the compliance with a seventh order polynomial looks thus:

$$\frac{1}{C_7} = 0.0334 - 0.2346\tau + 4.2615\tau^2 - 5.6959\tau^3 + \\ 3.5580\tau^4 - 1.1258\tau^5 + 0.1694\tau^6 - 0.0094\tau^7. \tag{5.4.2}$$

Note that for each order of approximation a new fit has to be done, and that all the coefficients may change. The approximate reciprocal compliance $1/C_7$ is shown in Figure 5.3.

Since the number of recorded data is not a whole power of 2, we used the fitted polynomial to calculate the Fourier coefficients, instead of interpolating the recorded data. The Fourier coefficients s_i and c_i can be calculated by separately integrating the fitted polynomial multiplied by the functions $\sin(2i\pi\omega t)$ and $\cos(2i\pi\omega t)$ over the period of the cycle

$$s_i = \int_0^T \frac{\sin(2i\pi\omega t)}{C_7} dt \quad i = 1, \ldots \tag{5.4.3a}$$

$$c_i = \int_0^T \frac{\cos(2i\pi\omega t)}{C_7} dt \quad i = 0, \ldots \tag{5.4.3b}$$

The Fourier approximation of the reciprocal compliance function is then obtained by summing the sine or cosine functions times the Fourier coefficients. At least one constant and five harmonics of the sine and cosine are needed to obtain a good approximation of the function

$$
\begin{aligned}
0.4916 &- 0.0597\cos[2\pi\omega t] + 0.7943\sin[2\pi\omega t]\\
&-0.3767\cos[4\pi\omega t] - 0.0944\sin[4\pi\omega t]\\
&+0.0886\cos[6\pi\omega t] - 0.0258\sin[6\pi\omega t]\\
&-0.0931\cos[8\pi\omega t] + 0.0500\sin[8\pi\omega t]\\
&-0.0084\cos[10\pi\omega t] - 0.0447\sin[10\pi\omega t].
\end{aligned} \tag{5.4.4}
$$

These approximations are shown in Figure 5.4

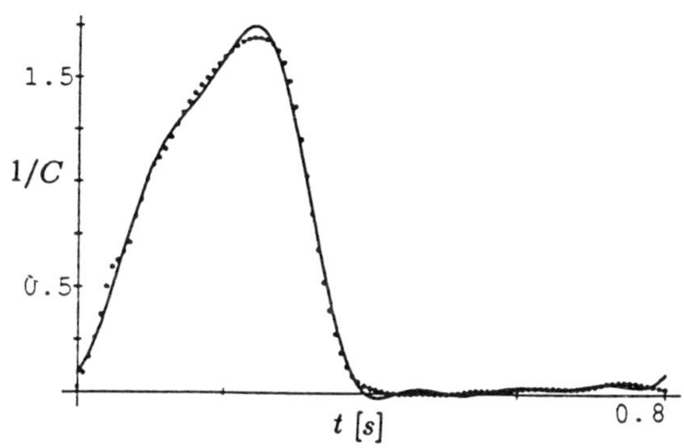

Fig. 5.4. Fourier approximation of the reciprocal of the ventricle compliance.

The calculation of the coefficients is quite tedious and the evaluation of the Fourier sum is time consuming in a simulation program, especially in our situation where at least ten geometric functions have to be evaluated for each step. The number of evaluations can be reduced by transforming a sum of a sine and cosine function with the same pulsation in a single sine function with phase angle ϕ_i, in the following fashion

$$a_i \sin(2i\pi\omega t + \phi_i) = s_i \sin(2i\pi\omega t) + c_i \cos(2i\pi\omega t) \tag{5.4.5}$$

where

$$a_i = \sqrt{(s_i^2 + c_i^2)} \quad \text{and} \quad \tan(\phi_i) = c_i/s_i. \tag{5.4.6a, b.}$$

This reduces the number of evaluations of geometrical functions per step by two, but one should be aware of the calculating errors in ϕ_i. As i increases, the coefficients s_i and c_i tend to zero, hence the error in $\tan(\phi_i)$ increases. Introducing erroneous phase angles results in changing the shape of the function, which should be avoided by all means.

One way to solve the problem is to use the polynomial function immediately. However a polynomial of order 7 is not a periodic function. This could be overcome by making the variable x a periodic function. The function $\tau(t)$ can be chosen as a saw-tooth function with period T

$$\tau(t) = t_{modT}, \tag{5.4.7}$$

which is shown in Figure 5.5.

Now the calculations are much easier to program since only seven instead of ten components have to be evaluated for each step (Table 5.2 below is an example).

In the next paragraph we will evaluate the vascular system using this approximation of the ventricular compliance.

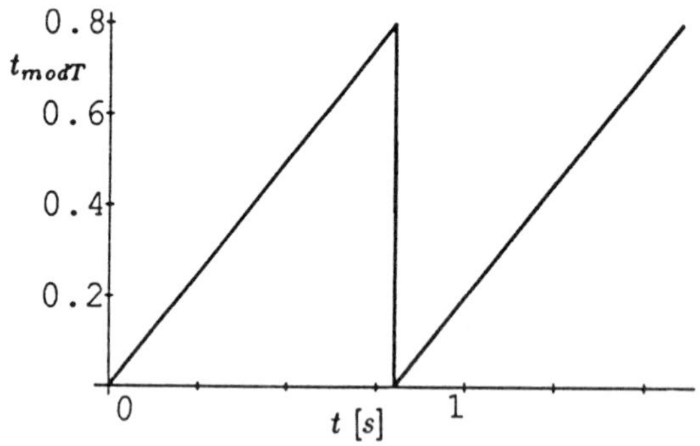

Fig. 5.5. Saw tooth function.

	50	...				;p
0.01	1000	gai	1013			;$R_{tra} f_{tra}$
	1005	sum	1042	1031		;$P_{tra} = P_{art} + R_{art} f_a$
	1011	sum	50	-1000	-1005	;$p - P_{tra} - R_{tra} f_{tra}$
0.0006	1012	att	1011			;$(p - P_{tra} - R_{tra} f_{tra})/L_{tra}$
0	1013	lmi	1012			;$f_{tra} = \max(\int (p - P_{tra} - R_{tra} f_{tra})/L_{tra}) dt, 0)$
0						;
1000						;
	1031	...				;$R_{art} f_a$
	1042	...				;P_{art}.

Table 5.2 List of instructions in TUTSIM-code to solve eqs 5.3.1 and 5.3.2

5.5 Evaluation of the vascular simulation

We have been solving the set of integral equations (5.3.1-5.3.10) numerically using the program TUTSIM. Notice that in eqs. (5.3.1), (5.3.3) and (5.3.8) the flows f_{tra}, f_{aor} and f_{ven} are not explicit functions, as they appear also in the integrand.

As we do not list the full set of instructions, the following example may illuminate all the other calculations. To solve (5.3.1) and (5.3.2) for example, the list of instructions V_0 is necessary (see Table 5.2). All the equations (5.3.1-10) can be similarly transformed into a list of instructions. As only one statement including one instruction can be carried out per line, complicated instructions should be broken down into a series of instructions. The ventricular pressure p for instance is calculated as v/C. During systole we use for $1/C$ the polynomial approximation as explained in the previous paragraph and we assume that during diastole $1/C$ is constant. The program TUTSIM sorts the statements, and during the run it calculates sequentially all the sorted statements. The output of a statement can be drawn immediately on screen or dumped into a file. The maximal number of output statements is confined to five. The output parameters are: the time t, the ventricular volume v, the ventricular flow f, the ventricular pressure p and the aortic pressure P_{aor}. We dumped these data in a data file with five columns from which two-dimensional plots were being extracted.

As before, using the program PASSAGE, from the dumped data we have plotted the graphs of the time-volume (t, v) (figure 5.6), the time-flow (t, f) (Figure 5.8), the time-ventricular pressure (t, p) (Figure 5.10), the volume-flow (v, f) (Figure 5.14), the volume-ventricular pressure (v, p) (Figure 5.12) as well as the flow-ventricular pressure (f, p) (Figure 5.16). We shall refer to these graphs as "simulated graphs" whereas the graphs of the time-volume (t, v), the time-flow (t, f), the time-pressure (t, p) and the time-aortic pressure (t, P_{aor}) of Figure 1.13, as well as the graphs of the volume-flow (v, f), volume-pressure (v, p) and the flow-pressure (f, p) of Figures 4.2 -4.4 will be called the "recorded graphs".

Although in the simulation we have omitted the atrial action, we choose the time t=0 at the point where the onset of the atrial contraction would have taken place. We now compare the simulated graphs to the recorded graphs.

First we compare the simulated time-volume and time-flow graphs of Figures 5.6 and 5.8 with the recorded ones in Figure 1.13. For convenience we have redrawn the latter in Figures 5.7 and 5.9. At the onset, the recorded volume increases (positive flow, Fig. 5.7) because the atrium is emptied into the ventricle. In the simulated graph (Fig. 5.6) the volume is constant (zero flow) because the atrial contraction is omitted. In both cases the maximal volume is called the EDV. When the aortic valve opens, ejection takes place and the volume decreases rapidly in both graphs (Figures 5.6 and 5.7) and the flow slows down.

When the atrio-ventricular valve opens, blood flows into the ventricle the pressure of which has dropped due to the relaxation and unwinding of the myocardial muscle fibres. The blood flows from the veins into the ventricle and, as a consequence, the ventricular volume increases. The inflow is not as fast as the ejection because of the small pressure difference between the veins and the ventricle. This suction is sufficient to maintain the blood supply in normal situations; normal meaning that the filling time is long enough, so that the pressure in the ventricle increases up to venous pressure.

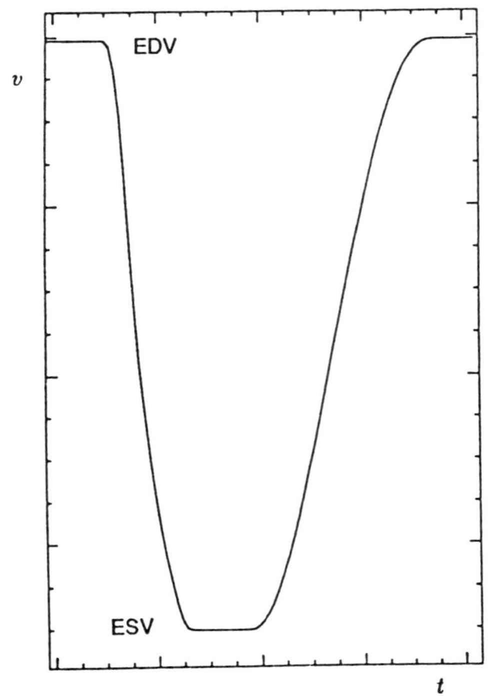

Fig. 5.6. Simulated time-volume graph of the left ventricle

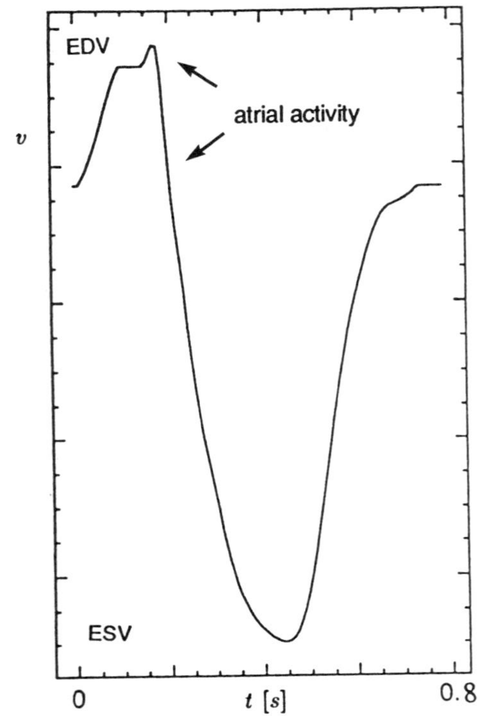

Fig. 5.7. Recorded time-volume graph of the left ventricle (from Ganong (1973)), from onset to .1 s the graph represents the atrial contraction, from .1s to .4s the ventricular ejection, from .4s to .8s the ventricular filling

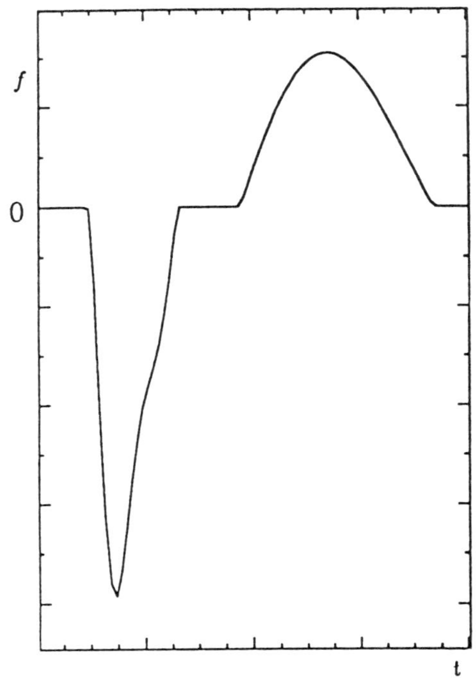

Fig. 5.8. Simulated time-flow graph of the left ventricle

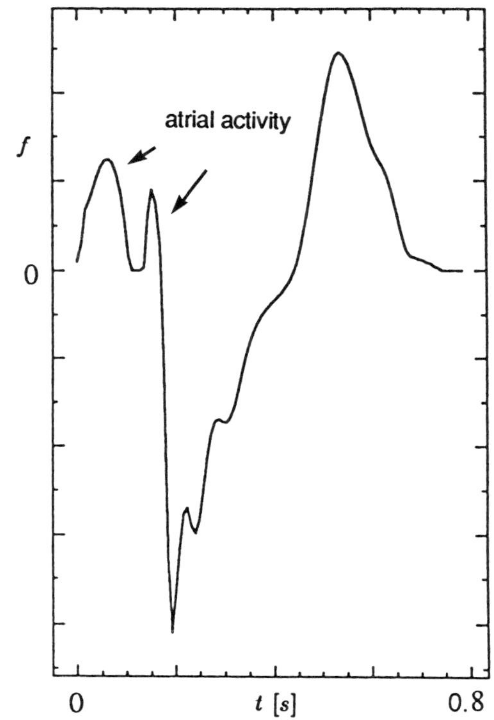

Fig. 5.9. Recorded time-flowgraph of the left ventricle (from Ganong (1973))

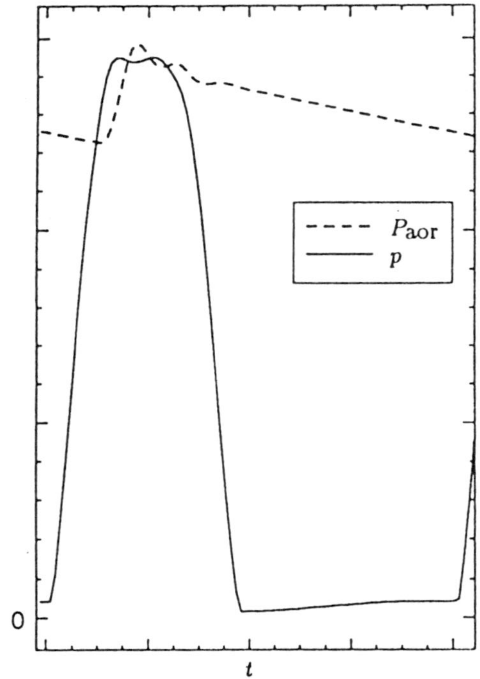

Fig. 5.10. Simulated time-pressure graphs of the left ventricle(——) and the aorta(- - -)

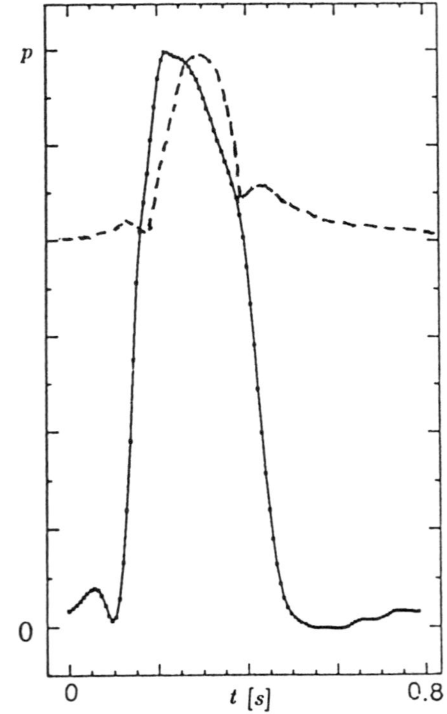

Fig. 5.11. Recorded time-pressure graph of the left ventricle (from Ganong(1973))

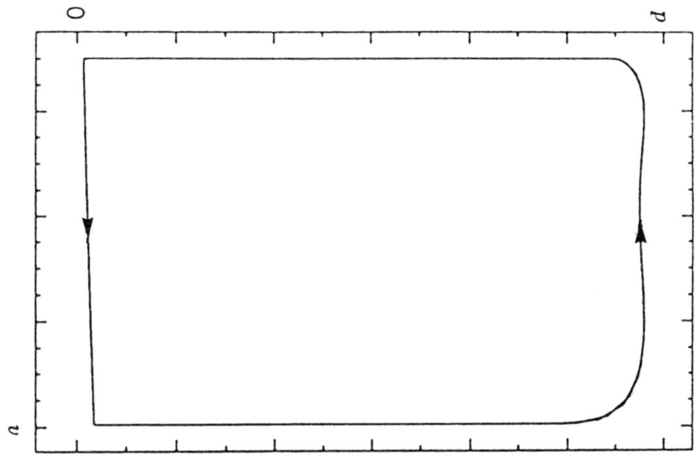

Fig. 5.12. Simulated volume-pressure graph of the left ventricle

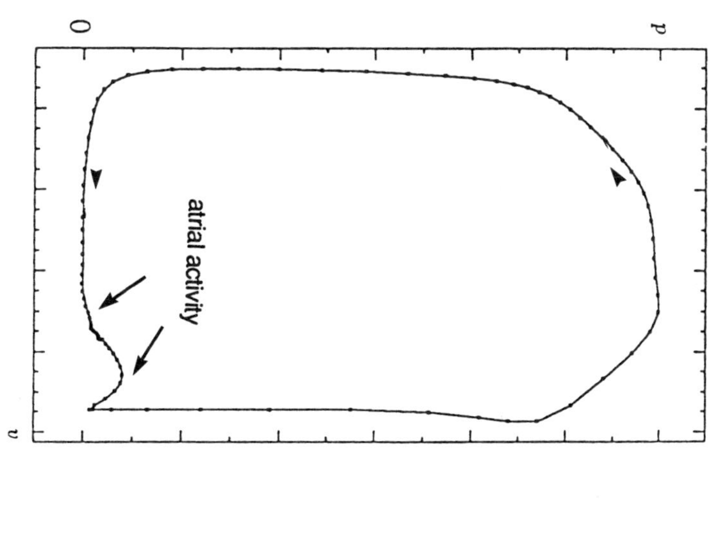

Fig. 5.13. Recorded volume-pressure graph of the left ventricle

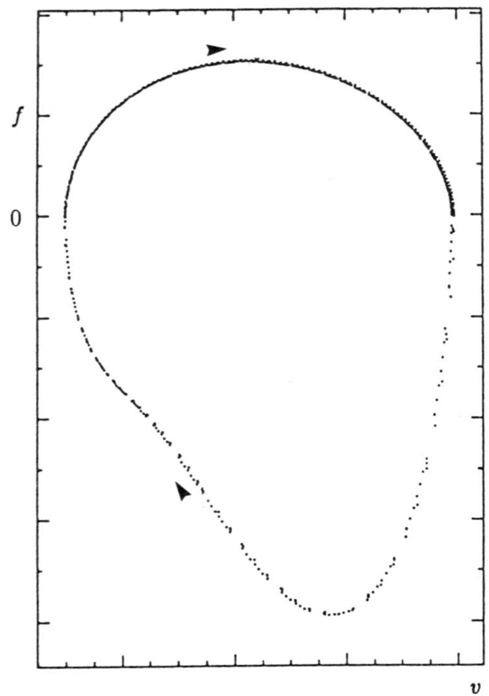

Fig. 5.14. Simulated volume-flow graph of the left ventricle

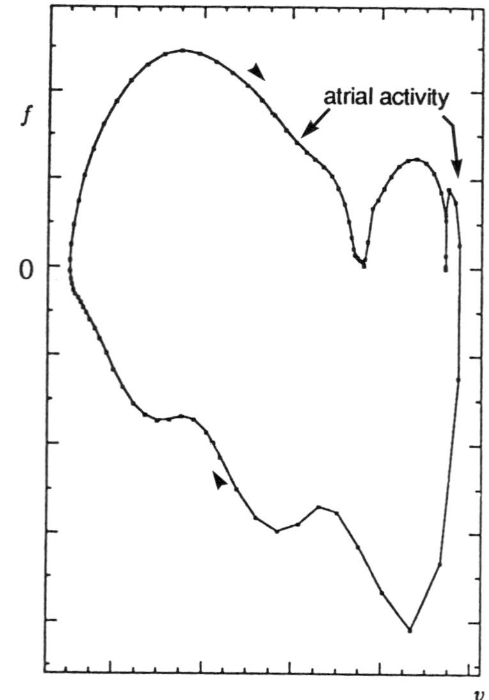

Fig. 5.15. Recorded volume-flow graph of the left ventricle

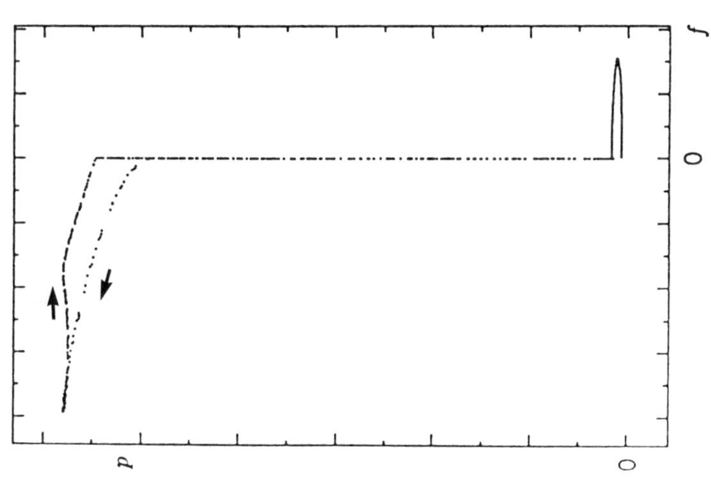

Fig. 5.16. Simulated flow-pressure graph of the left ventricle

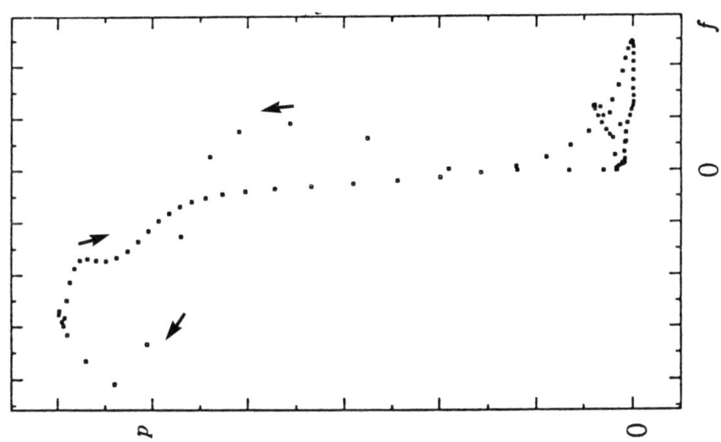

Fig. 5.17. Recorded flow pressure-graph of the left ventricle

With higher heart rate the diastole shortens. This tends to a smaller filling of the ventricle. An additional filling by the atrium is in order to maintain or increase the performance of the ventricle. In normal situations the atrial contraction supplies only a small fraction of the blood. To obtain a comparable volume increase in the simulated model we had to elongate the inflow time, so that the simulated graph is broader than the recorded one during diastole. In the recorded graph there is a supplementary bulge in the flow graph, due to atrial action.

Secondly, the time-flow diagrams are shown in Figure 5.8-9. The recorded plot starts also at the onset of atrial contraction. During the diastolic phase from $t = 0.00s$ to $0.15s$, the recorded flow (Fig. 5.9) shows two bulges, the first is due to the atrial contraction, the second does not turn up in every publication on the subject. We will comment on this as we discuss the (f, p)-plots. In the simulations we skipped this initial rise (atrial contraction). A steep decrease is recorded afterwards (0.15 - 0.2s), followed by a slower return to zero (0.2 - 0.4s). The recorded flow shows an oscillating behaviour during this return to zero, which probably is the result of pressure or resistance changes over the aortic valve. These oscillations are however not simulated. From zero the flow becomes positive as the AV-valve opens (0.4-0.8s). The maximal inflow rate in the ventricle is about half of the value of the maximal outflow rate. Since the surface under the curve is a measure for the volume, the simulated inflow buldge has to be broader than the recorded one, because the atrial contraction is omitted in the simulation.

Thirdly, we compare the recorded (Figure 5.11) and the simulated (Figure 10) pressure diagrams. The ventricular pressure p is drawn as an uninterrupted line while the aortic and venous pressures are drawn by a broken line. They occur at, respectively, a high and a low level. The simulated pressure matches the recorded one very well. At the onset of the contraction both pressures rise until the pressure in the ventricle is equal to the aortic pressure. Then the aortic valve opens. From this moment on the ventricular volume decreases, but notice that the ventricular pressure still rises due to the rising reciprocal compliance. As the blood is ejected into the aorta the aortal pressure rises as well. The inertance of the blood and the filling of the arterial system maintains the outflow even if the aortal pressure exceeds the ventricular pressure. When the aortic valve closes the aortic pressure falls steeply in both simulated and recorded graphs, and attains a level lower than the venous pressure. During diastole the ventricular pressure increases slightly due to the volume increase, while the venous pressure decreases. In the recorded graph a supplementary bulge is seen during the atrial contraction. During ventricular diastole and atrial systole the aortic pressure decreases constantly until the aortic valve opens and the aortic and diastolic pressures equalize. The maximal aortic pressure is called the *systolic pressure*.

In the following graphs the time is eliminated. Each point represents a time interval of 8 ms. The (v, p)-plots (Fig. 5.12-13) show two jumps from low to high pressure and vice versa at points where the ventricular volume is constant. The simulated (Fig. 5.12) and the recorded (Fig. 5.13) graphs are very similar. In the latter a supplementary atrial action is present. At the high pressure level, both simulated and recorded graphs show a fast emptying while the pressure is still rising, followed by a slower volume decrease while the pressure is decreasing. In the recorded system the peek systolic pressure (PSP) is attained closer to the EDV than to the ESV. While in the simulated graph, the PSP appears closer to the ESV. This is due to the simplification of the vascular system.

Also the simulated (Fig. 5.14) and recorded (Fig. 5.15) volume-flow plots are very much alike. From the EDV the flow falls to a large negative value. Then the flow decreases up to half the minimal flow value where there is a notch. Then the flow decreases to zero while the atrial valve closes. When the AV-valve opens the inflow into the ventricle increases and it decreases again when the ventricular pressure exceeds the venous pressure. Again, the graph shows a supplementary structure due to atrial contraction.

The simulated (Fig. 5.16) and the recorded (Fig. 5.17) flow-pressure diagrams seem to differ a lot in the steeply increasing part of the pressure. This is due to recorded ventricular volume changes. As already stated, the heart specialists differ as to the interpretation of this phenomenon. Some explain the recorded flow as an artefact due to measuring methods, whereas other investigators insist on the existence of a flow during contraction. For simplicity we accept the first interpretation. The small loop in the low pressure range is due to the contraction of the atrium.

Chapter 6 Three-dimensional Catastrophe Model for the Ventricle

6.1 The matching-simulation method

We now try to replace the left ventricle, which has been simulated in the previous chapter by a variable compliance, by a catastrophe machine. We already observed that in the recorded volume-pressure plots there are two pressure jumps at almost constant volume. Thus, to obtain a graph with two jumps, the catastrophe type of that machine should be at least of order three. Hence, we try the simplest one, viz. the cusp catastrophe with equation (2.4). As explained in Chapter 3, we can associate with the cusp catastrophe a dynamic system which is given by equations 3.3.3 and 3.3.5. For convenience we renumber the latter equations as

$$\dot{a}_1 = \frac{-1}{K}x, \qquad (6.1.1)$$

and

$$\dot{x} = -K(x^3 - a_2x - a_1). \qquad (6.1.2)$$

For constant $a_2 > 0$, Figure 3.4 shows a plot with two jumps in the (a_1, x)-plane. One might be led to think that the recorded volume-pressure plot from Figure 5.13 should be modelled onto the graph of Figure 3.4, meaning that the state x should play the role of the ventricular pressure p and that a_1 could be interpreted as the ventricular volume. This would imply that, due to (6.1.1), the derivative of the ventricular volume $\dot{v}$ is equal to $-p/K$, i.e. to the ventricular flow f. However, a linear (f,p)-plot contradicts the findings both in the recorded and simulated vascular system. It can be seen from Figure 5.17 that the recorded flow and pressure graphs are not straight lines at all. This does not mean that the cusp catastrophe may not be a good model, but only that this specific identification between a_1 and v is not the right one. Another remark is that the (a_1, x)-plot is symmetrical over its origin. If x is identified with the ventricular pressure p, then the values of the venous part should be adapted to obtain an acceptable flow during diastole.

This is an example of how one may try to *match* a catastrophe type to a recorded dynamical system. It is appropriate to devote a few remarks here to the method which underlies this monography. In fact, the following remarks may be relevant in connection with *any* attempt aimed at verifying the presence of a catastrophe-type dynamic in Biology by way of comparing a simulated process with a recorded one. Indeed, verifications of this kind take place in fields that do not always allow the construction of strictly deductive-mathematical models as in classical physics and dynamics. In fact, verifications of this kind are common in sciences that are interdisciplinary between Physics and Biology. That is to say, they concern processes for which one cannot derive, in terms of certain measurable quantities (forces, energy, time, etc.) that are seen on the process, equations that supposedly hold for that process, and that, in turn, can be tested experimentally.

Due to the abstract mathematical representation of catastrophe-types in rational, standard or minimal form, one may not expect that its state- and control parameters may be directly matched to quantities that are measurable in terms of known physical dimensions. For instance, the representation may need a diffeomorphic transformation of its (state and control) parameters before they correspond to experimental quantities. Conversely but just as important, one may not even be able to formalize certain experimental quantities into one arithmetical connective model together with the other experimental quantities, in order to derive a type. In the present case for instance, suppose for a moment that the ventricular suction pump is a realization of the cusp catastrophe type, describable in terms of the measurable quantities p, v, f and time t'. Then the standard (rational) representation (Chapter 2, formulas (2.4-7) may be "removed" from that realization by a diffeomorphism, i.e. by an invertible C^∞ transformation from the (v, f, p, t')-system to the (a_1, a_2, x, t)-system. Only such transformations are known to preserve the jump features of the (graphs involving the) state variables (in our case p and x), whereas they may deform the other features of the graphs in an unrecognizable manner. Since there exist continuously many diffeomorphisms one might rightly say that finding the right one is like finding a needle in a haystack. This is precisely the reason why inveterate deductivists even among applied mathematicians, take a skeptical look at results that match a catastrophe type to experimental data in this way, and then proclaim that catastrophe theory works. Indeed, results obtained in this way are often ambiguous, and at best approximate, meaning that the mathematical model may describe the process more or less tightly, up to a certain inscrutible transformation, i.e. in a manner which falls outside the realm of deductive control. And this is not what mathematicians of the classical schools like.

W. Kuyk designed what may be called the *matching-simulation method*, which this book exemplifies as follows. One first *matches* a known catastrophe type to a recorded process that one intends to describe, by determining the parameters on record that may correspond to the state- and control parameters of a conjectured catastrophe model. In the previous chapter we have proceeded to do so in the ventricular pump case by finding p to jump periodically between systolic and diastolic states (in geometric jargon: between the attractor sheets 'systole' and 'diastole'). This p then was conjectured to 'correspond' to the variable x of the cusp type. Here matching includes the arduous task of trying to map well-chosen (but not necessarily all) control variables to (not necessarily all) recorded parameters or quantities, and then checking whether things work by comparing the recorded graphs with computed ones. In fact, in what follows we try to *simulate* the recorded graphs starting from the theoretical model, by identifying recorded parameters with mathematical combinations of control parameters of the cusp type. There is no procedure here other than the trial and error method and one is led by the hope that the theoretical unfolding of the cusp type that is used, matches the unfolding presented by the natural process.

The words "first match" and "not all parameters" in the previous paragraph refer to the fact that certain recorded parameters that are judged of 'first order' are included, whereas other 'second and higher order' parameters are temporarily deleted from the simulation. Up till now, for the ventricular pump we included the ventricular pressure, volume and flow, but not ventricular stretching and contractility. This means that we deal with a 'dull'ventricle, i.e. a myocardium that does not readily stretch upon receiving an unexpected increased inflow (e.g. because it is old and worn) and that knows no 'contractile states'; a myocardium, in short, that possesses no degrees of muscular elasticity and resilience.

6.2 Matching the first order parameters

We have already noted that the normal parameter a_1, which is primarily responsible for forcing $x = p$ to catastrophically jump, is not to be identified with v. The same method shows, a fortiori, that it should not be identified with $f = \dot{v}$. Thus, we call a_1 the ***dynamic parameter*** of our system, since it controls p via equations 3.3.3 and 3.3.5. It is an example of a theoretical parameter that has no apparent counterpart on the record. Experimentations lead us in putting $a_2 = 1$ and $K = 5$, causing catastrophic jumps to take place close to $a_1 = \pm\sqrt{1/3}$. Here we apply the following argument. It is readily seen that the value of K determines the steepness of the jumps in the (a_1, x)-graph as well as in the (x, t)-graph. The larger K, the jerkier the graph is.

Figure 6.1 shows (a_1, x)-graphs for different values of K between 0.2 and 5 while a_2 is kept constant to 1. The broken line is the graph of equation (2.4).

On the other hand, the larger K, the closer the trajectories approach the upper and lower sheet of the broken line. For a value of K equal to 5 we obtain a graph that is acceptable as a candidate for the catastrophic jumps. Due to (3.3.8) the value of K also controls the period of the system; i.e. further increasing K would cause a period for the system, away from an acceptable heart period. More precisely, when a_2 and K are respectively equal to 1 and 5, the period of the system is equal to 11.5 s. This is far too large for a simulation of the heart beat

Using the other control variable a_2 we may change the periodicity of the signal (cf. Chapter 3), but this causes a change in the maximal pressure amounting to $x_{\max} = 2\sqrt{a_2/3}$. In Figure 6.2 we plot the maximal value of x $(x_{\max})$ as a function of a_2 together with the curve $x = \sqrt{a_2/3}$, showing that in the range $a_2 = 0.2$ to 4, the simulated maximal pressure lies on the curve. *Notice also the similarity between Figure 6.1 and Figure 1.5, i.e. the relationship between EDV and performance (Frank Starling Law)*

Now for the period T, which is also a function of a_2. In Figure 6.3 the simulated period T ranges from 5s to 40s as a_2 ranges from 0.2 to 4. For $a_1 = 1$ and $K = 5$ we fit the line

$$T = 4.00s + 7.96sa_2$$

through the data points. However, varying a_2 we are not able to obtain a system with period 0.8 s for $K = 5$ and a positive value of a_2 (note that the value of a_2 should be positive if one wants a jump in the (a_1, x)-plane). In order to decrease the period of the system we introduce a time scaling parameter S into equations (6.1.1-2), to obtain

$$\dot{a}_1 = \frac{-S}{K}x \tag{6.2.1}$$

and

$$\dot{x} = -SK(x^3 - a_2x - a_1). \tag{6.2.2}$$

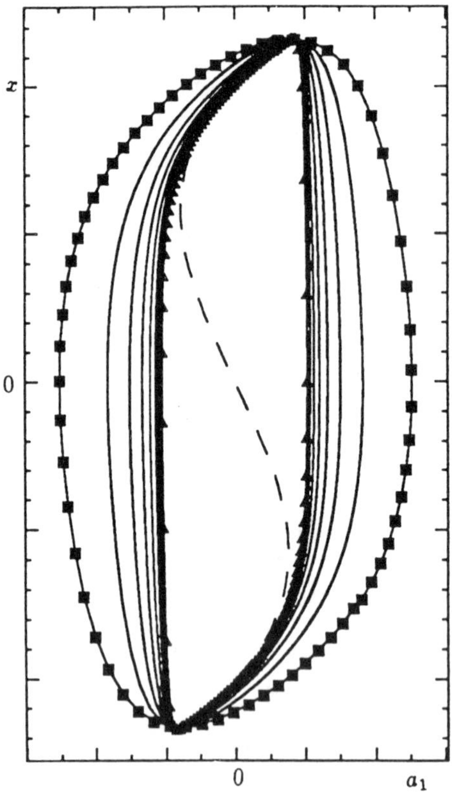

Fig. 6.1. (a_1, x)-graphs for different values of K between .2 and 5 while $a_2 = 1$. The curve for $K = .2$ is shown as an uninterrupted line with rectangles, while the curve for $K = 5$ is marked with triangles. The broken line shows the equation $x^3 - a_2 x - a_1 = 0$

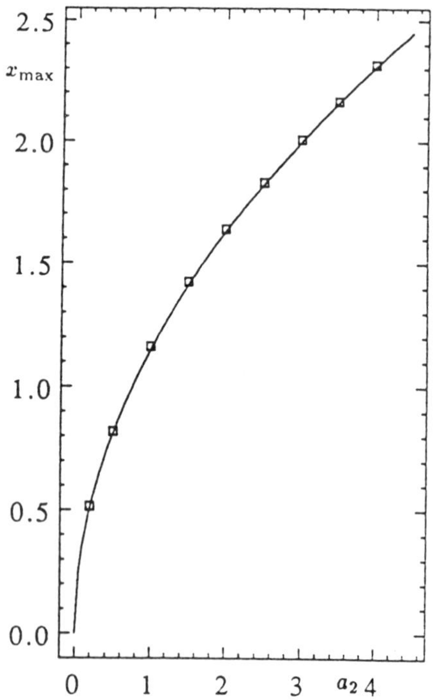

Fig. 6.2. x_{max} for different values of a_2 between .2 and 4 while $K = 5$ and the curve $x = 2\sqrt{a_2/3}$.

With this scaling the shape of the (a_1, x)- graph is not modified, only the time graphs show a scaling in the t-direction.

Setting S to 14.5 we obtain a oscillator (by Eq. 6.2.1) with period 0.8 s ($\approx 11.5/14.5$). The state parameter ranges from $-2\sqrt{a_2/3}$ to $2\sqrt{a_2/3}$ whereas the recorded pressure is normally positive. The minimal value is $x_{\text{min}} = -2\sqrt{a_2/3}$. To obtain a more realistic pressure volume graph, it can be shifted up, by adding a constant to the ventricular pressure.

The technique used is to sample and hold the value of x when $\dot{x}$ changes from $-$ to $+$. The minimal value x_{min} of x is negative. The value of $\dot{x}$ changes from $-$ to $+$ at the end of the catastrophic fall when x starts to increase. The value of the constant c is defined as $c = -x_{\text{min}}$. The value of c is constant during one cycle. Only at the end of the catastrophic fall can this constant change if the parameter a_2 changes.

With these choices, we now compute and produce all the time graphs and all the time free graphs that we can, and compare them to the recorded graphs and, where indicated, to the graphs after Rideout. The computed time graphs are shown in Figures 6.4-7, and the other graphs are shown in Figures 6.8-10.

Now, the inflow time of 0.25 s is comparable to the recorded inflow time. In the simulation after Rideout of the previous chapter, the inflow time turned out slightly elongated as compared to the record, due to a small pressure difference between the venous and the ventricular pressure during diastole. We recall that the atrial action was not taken into account.

As for the outflow, during ventricular systole it is shorter than the recorded one but is as short as in the simulation after Rideout. The catastrophe simulation does not show the notch in the flow as well. In Figure 6.6 the ventricular pressure p is drawn as an uninterrupted line while the aortic and venous pressures are represented by broken lines. The ventricular pressure rises steeply from $p = \sqrt{a_2/3} + c$ to its maximal value $p = 2\sqrt{a_2/3} + c$. When the ventricular pressure reaches the level of the aortic pressure the aortic valve opens and blood is ejected into the aorta. Thus the aortic pressure increases and even exceeds the ventricular pressure. From its maximal value the ventricular pressure decreases monotonically from a value $p = 2\sqrt{a_2/3} + c$ to $p = \sqrt{a_2/3} + c$, due to the decrease of a_1. The aortic valve closes as soon as the ventricular flow has become zero. Counting from $p = \sqrt{a_2/3} + c$ the ventricular pressure falls steeply to its minimal value $p = -2\sqrt{a_2/3} + c$. If this level is lower than the venous pressure, than blood flows into the ventricle. Due to the increase of a_1 the ventricular pressure rises up to $p = -\sqrt{a_2/3} + c$, and the cycle restarts again. The pressure increase during ventricular diastole is steeper than the recorded one. During ventricular systole, the simulated pressure matches the recorded one very well (see Figure 5.11). The resulting pressure pattern is periodical with period T =0.8 s. The period can easily be modified by changing the time scaling parameter S, as we have seen. Figure 6.7 shows the dynamic parameter time plot. The (a_1, t)-graph has the shape of a triangle wave which reaches its extreme values for $x = 0$. The dynamic parameter a_1 forces the catastrophic jump of x which is shown in Figure 6.8 in the (a_1, x)-plot. In this plot as well as in the following one the time is eliminated. The jumps occur when the value of the dynamic parameter is close to $a_1 = \pm\sqrt{4a_2/27}$. The steepness of the jump can be modified by setting the parameter K. This graph shows a point symmetry with respect to the origin, due to the symmetry of equation (2.4).

The volume-flow (v, f) plot is shown in Figure 6.8. Note that in the simulated graph, the flow falls steeply to a large negative value at the EDV. In the following slow decrease to zero, one cannot observe a notch (see Figure 5.15). During ventricular diastole the flow rises up to half the absolute value of the outflow. The flow-pressure (f, p)-plots (see Figure 6.9) obtained from the catastrophy model and the simulation after Rideout (see Figure 5.17) are very much alike.

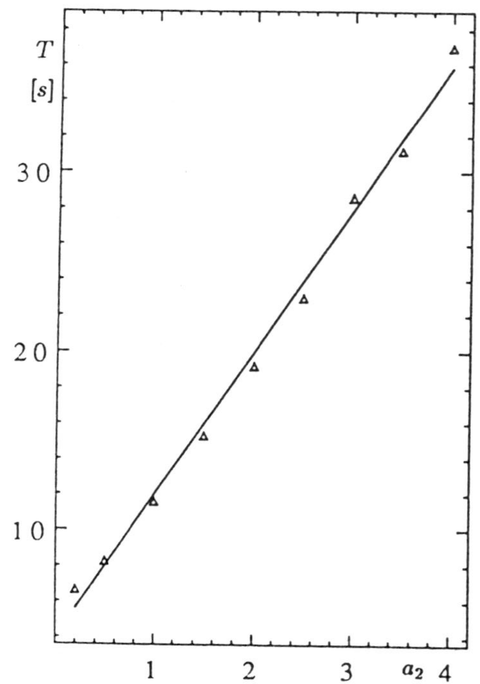

Fig. 6.3. T for different values of a_2 between 0.2 and 4 while $K = 5$ and the fitted line through the data points.

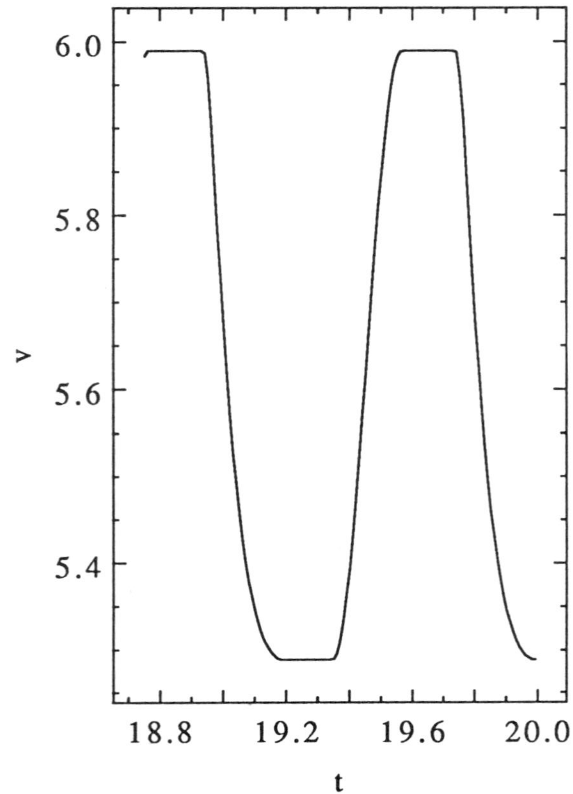

Fig. 6.4. (t, v)-graph from the cusp model with $a_2 = 1$

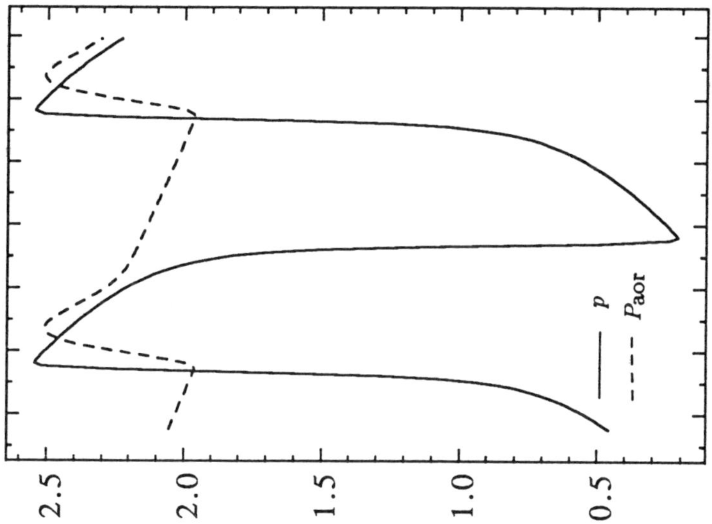

Fig. 6.6. (t, p)-graph from the cusp model with $a_2 = 1$

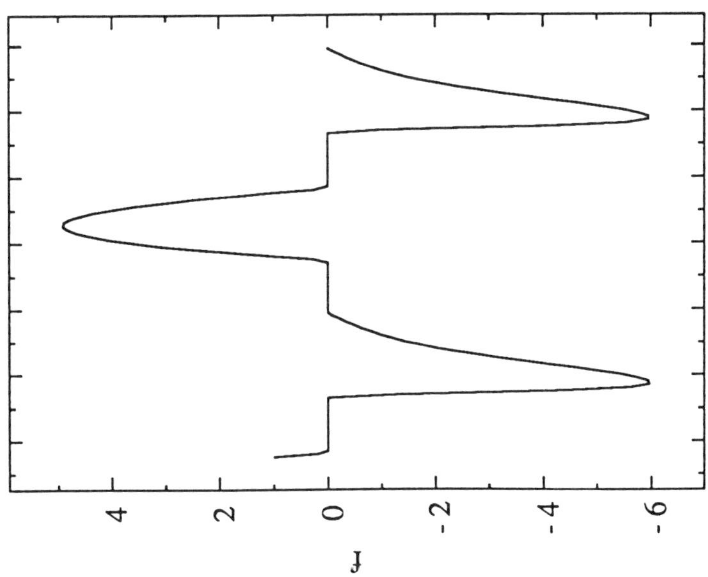

Fig. 6.5. (t, f)-graph from the cusp model with $a_2 = 1$

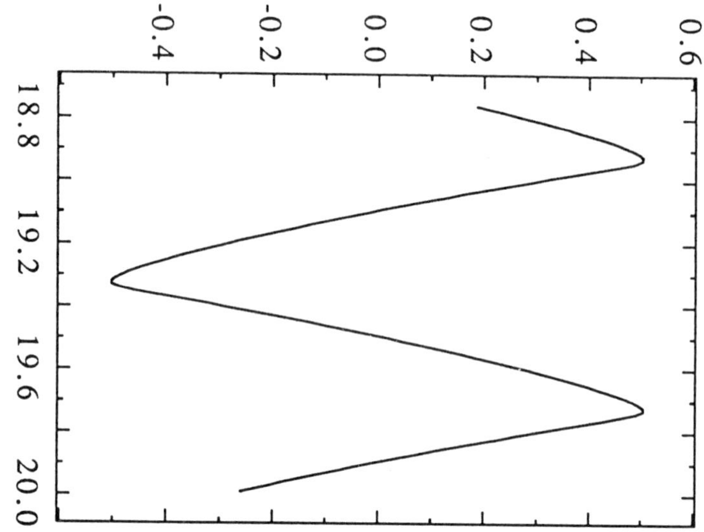

Fig. 6.7. (t, a_1)-graph from the cusp model with $a_2 = 1$

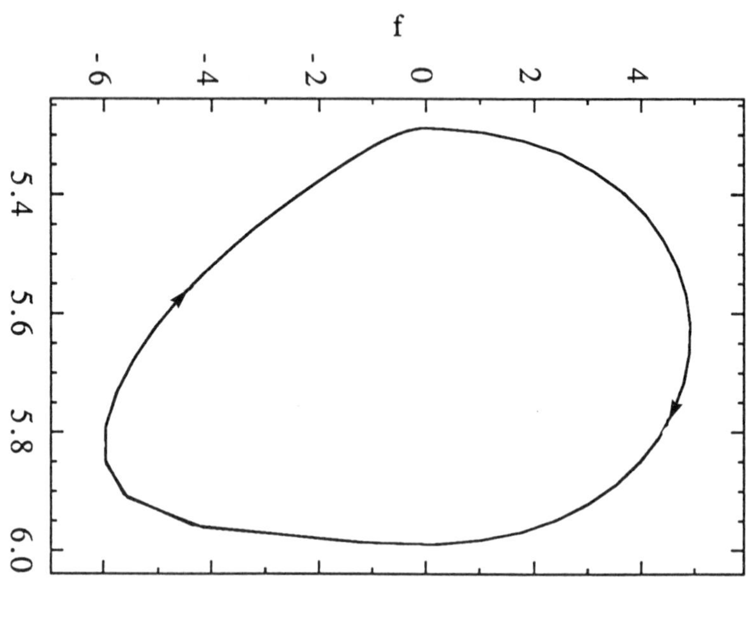

Fig. 6.8. (v, f)-graph from the cusp model with $a_2 = 1$

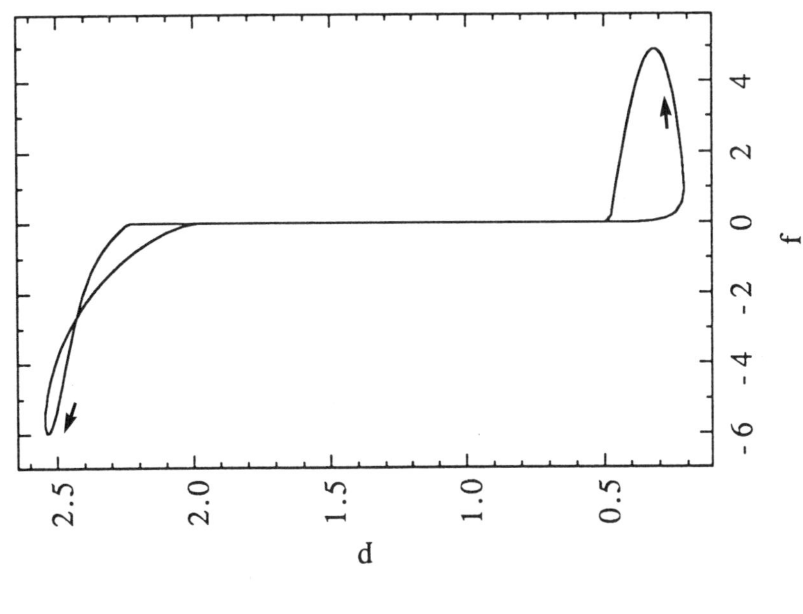

Fig. 6.9. (f,p)-graph from the cusp model with $a_2 = 1$

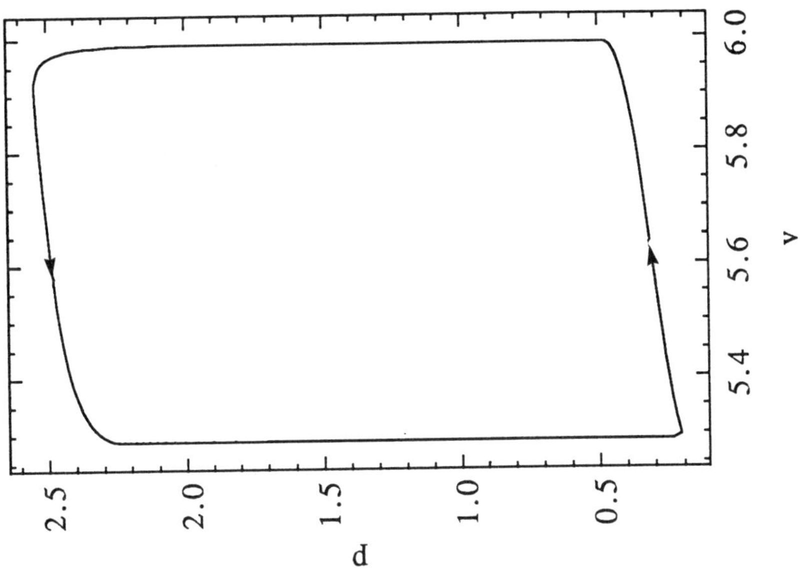

Fig. 6.10. (v,p)-graph from the cusp model with $a_2 = 1$

The volume-pressure plot is shown in Figure 6.10. This plot is very similar to the recorded one (see Figure 5.13). However, the volume-pressure plot is symmetrical over its centre, the peek systolic pressure (PSP) is reached when the ventricular volume is close to the EDV and the ventricular pressure is not related to the EDV.

To obtain a good simulation of the heart the "Frank-Starling" relationship should be included. This can be done in the following way.

From Figure 6.2 we see that the maximal pressure is related to the value of a_2 by $p_{\max} = 2\sqrt{a_2/3}+c$. This means $x_{\max}$ is steadily increasing as a function of a_2. To obtain a relation between the ventricular volume v and the pressure we suggest identification of the ventricular volume v with a_2. We have to modify the fast equation (6.2.1) to become

$$\dot{x} = -SK(x^3 - vx - a_1). \tag{6.2.3}$$

The relationship (6.2.3) is remarkable, since it says that the pressure depends on the ventricular volume v, reminding us of the well-known "Frank-Starling Law" of the heart.

To better see the meaning of this relationship we list the obtained time graphs (Figures 6.2.11-14) as well as the timefree graphs (Figures 6.2.15-17). These graphs compare very well to the graphs where a_2 is kept constant and equal to 1. However, after (6.2.3) the ventricular pressure p and the end diastolic ventricular volume EDV are interrelated. The maximal ventricular pressure will be close to $x_{\max} + c = 2\sqrt{EDV/3} + c$. This value introduces into the simulation a relationship between ventricular pressure and ventricular volume similar to "Frank-Starling".

Another remarkable result of the identification of a_2 and v is, that the system does not have an exact periodic behaviour any more. Due to equations 6.2.1 and 6.2.3, the period depends on the ventricular pressure, which in its turn depends on the ventricular volume and thus on the entire cardio-vascular system. Thus, changes in that system affect not only the ventricular pressure, but also the rhythm of the beating. This phenomenon has been observed before, and this is the first time that it is being simulated from a model.

The peak systolic pressure (PSP) is reached when the ventricular volume is close to the EDV, and the ventricular pressure is decreasing from that point onward during the rest of the systole. Notice also that the (v,p)-graph is not point-symmetrical with respect to its centre. This is due to the fact that the value of a_2 is not the same. During the catastrophic fall the value of $v = ESV$ and during the catastrophic rise of the value op $v = EDV$. Moreover, it makes the catastrophe points $x = -\sqrt{EDV/3}$ and $x = +\sqrt{ESV/3}$ to be fixed and dependent only on the end diastolic ventricular volume EDV and end systolic ventricular volume ESV.

6.3 Higher order models

The cusp catastrophe model of the cardio-vascular system presented thus far includes the ventricular volume as the only external parameter. The peak systolic pressure (PSP) is also reached when the ventricular volume is maximal. It is clear that not all the features of the complex cardiac system can be explained with this simple model of a manifold whose level curves (i.e. intersections of the catastrophe manifold with planes defined by taking p, a_1 or a_2 to be constant) are generically of degree three; the bifurcating cusp being of type (3,2) (cf. Chapter 2). We have already shown that the degree of the cusp

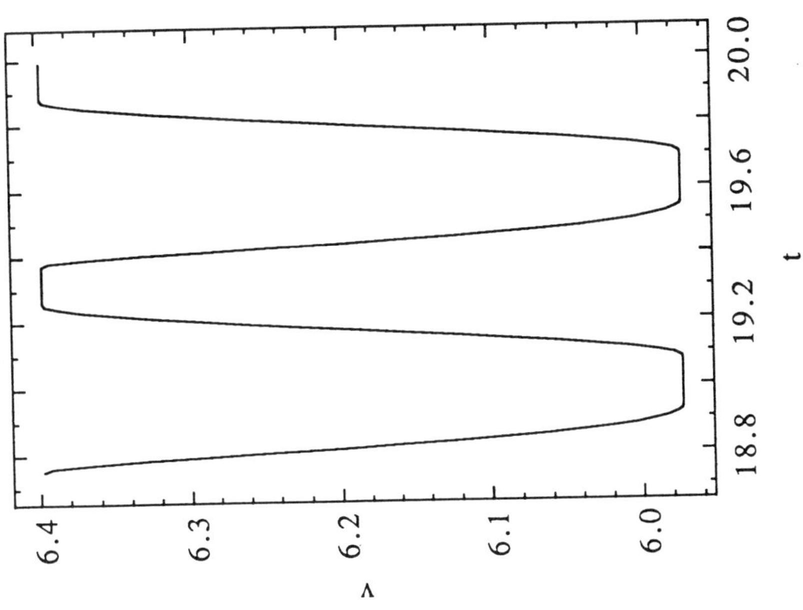

Fig. 6.11. (t, v)-graph from the cusp model with $a_2 = v$

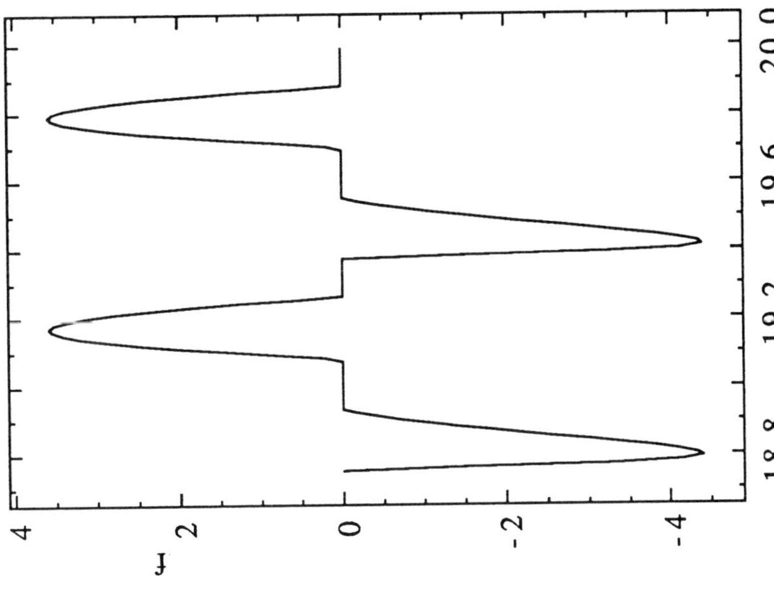

Fig. 6.12. (t, f)-graph from the cusp model with $a_2 = v$

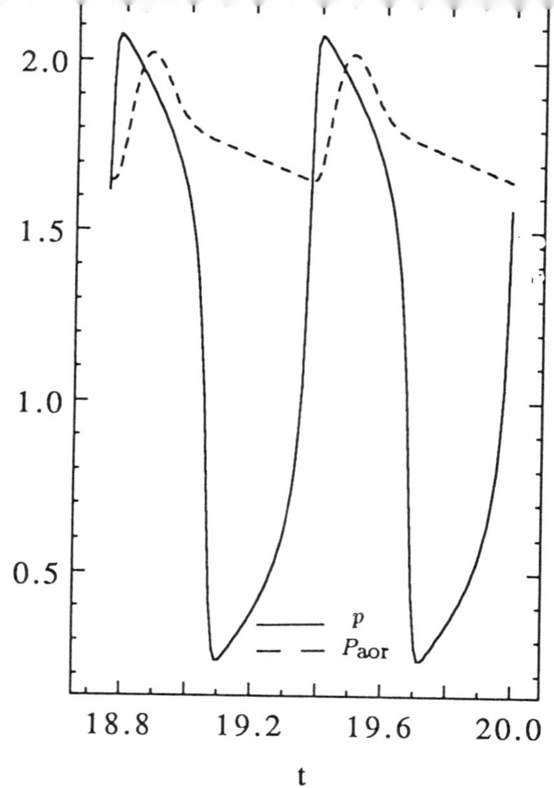

Fig. 6.13. (t, p)-graph from the cusp model with $a_2 = v$

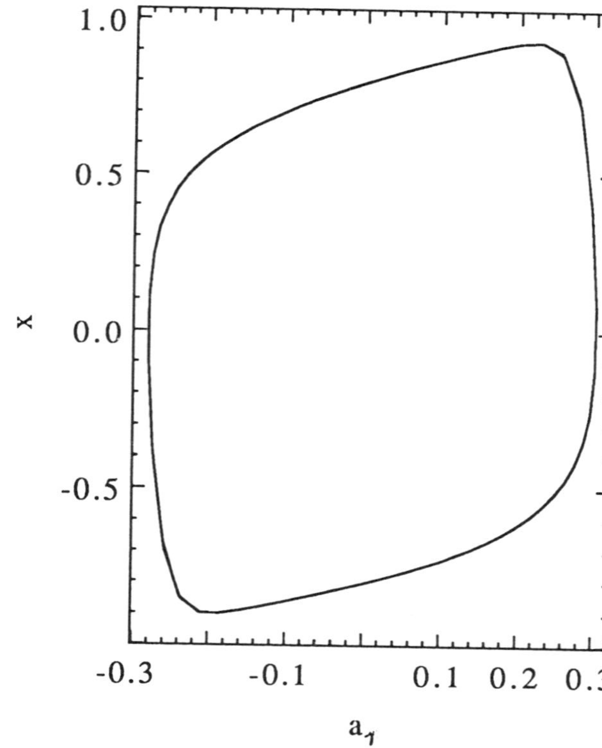

Fig. 6.14. (t, a_1)-graph from the cusp model with $a_2 = v$

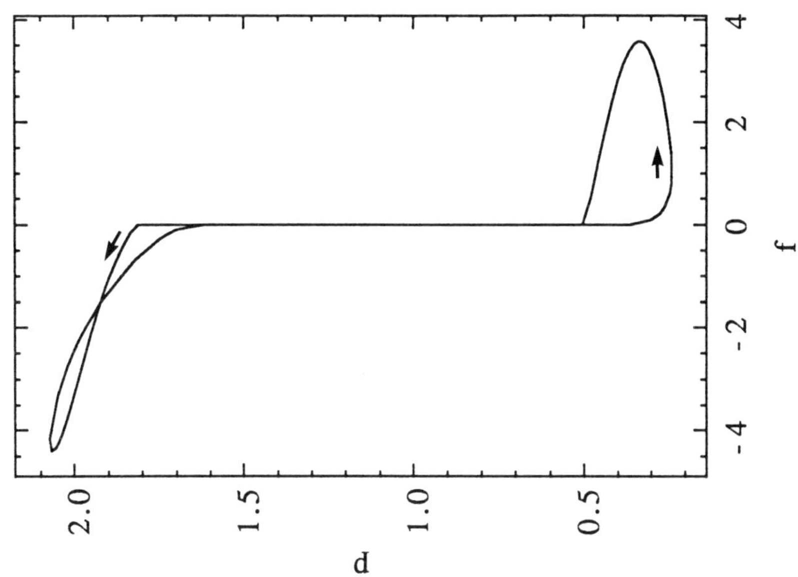

Fig. 6.16. (f,p)-graph from the cusp model with $a_2 = v$

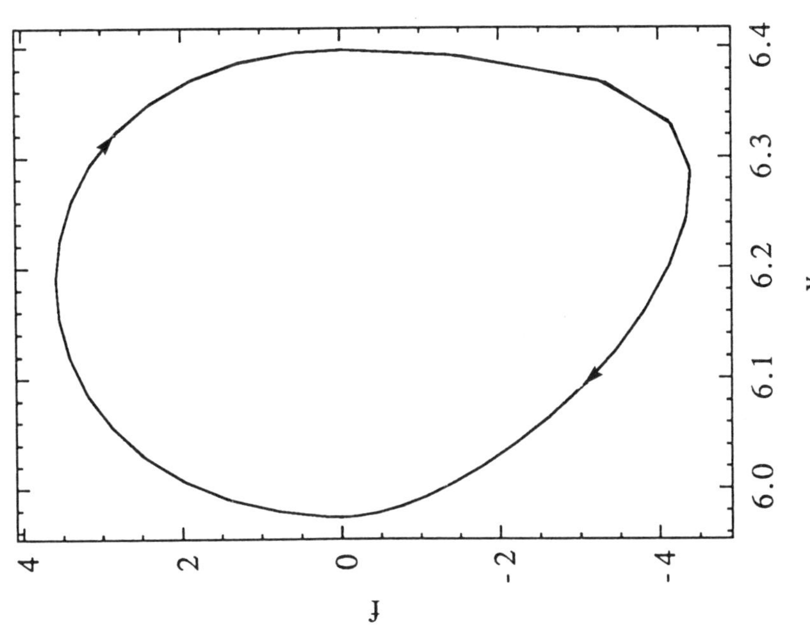

Fig. 6.15. (v,f)-graph from the cusp model with $a_2 = v$

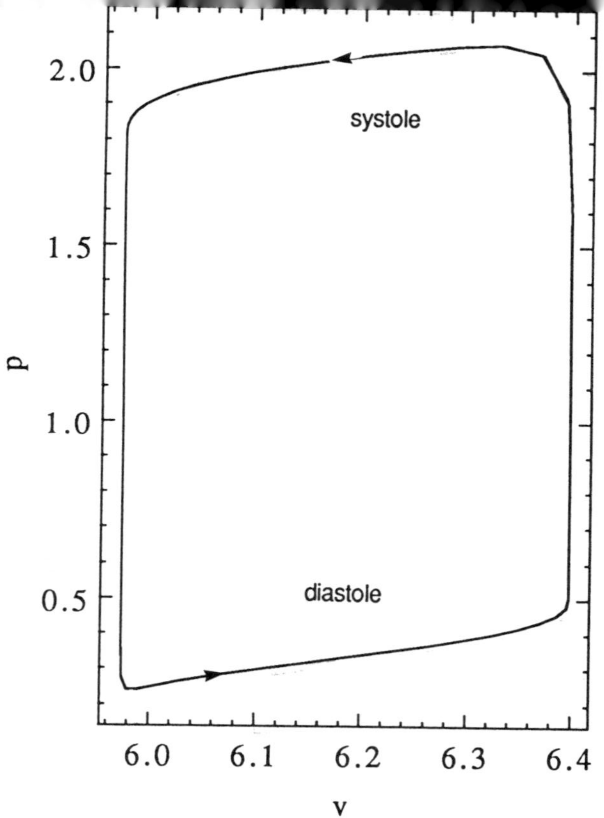

Fig. 6.17. (v, p)-graph from the cusp model with $a_2 = v$

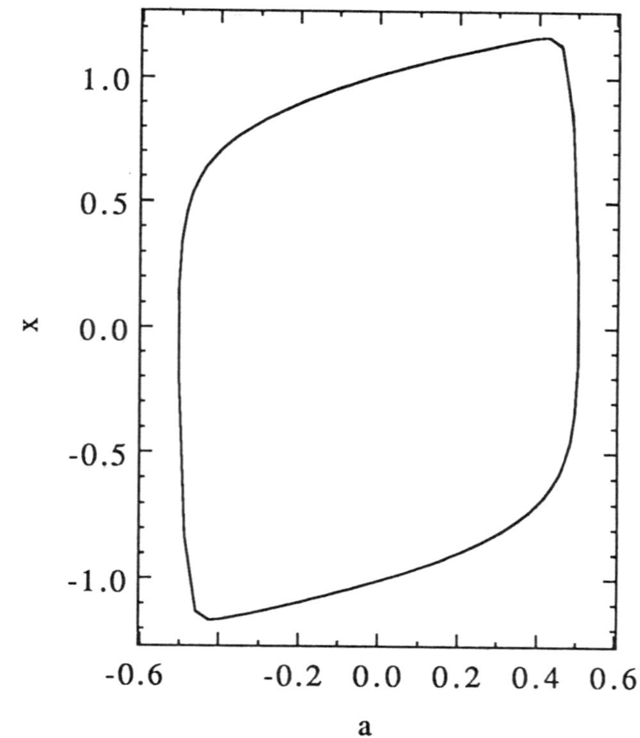

Fig. 6.18. Shifted (a, x)-graph from the cusp model with $a_2 = v$

in the control plane can be set to any odd number n (formula (2.7)), and that one can 'generalize' the present van der Pol oscillator to admit level curves of degree n (Chapter 3, paragraph 3.4) simply by replacing x^3 with x^n. Thus, the oscillator formula (6.1.3) takes the shape

$$\dot{x} = -SK(x^n - vx - a_1), \tag{6.3.1}$$

and equation (6.1.1) remains the same but is renumbered for convenience as

$$\dot{a}_1 = (\frac{-S}{K})x. \tag{6.3.2}$$

Let it be remarked that this increase from third to fifth degree introduces only slight changes in the graphs of the slow manifolds (which we do not draw here), since it preserves the nature of the oscillator as an oscillator on a cusp surface which looks pretty much the same as the one defined by the formulas (2.2-3), and which is only a partial unfolding of x^5. However, the higher degree puts us in a position to introduce two new parameters into the equations as follows. From now on, instead of applying the unfolding of x^3 we consider the full unfolding of x^5 by setting in (2.1) and (2.2) the integer n equal to 5. Hence, we not only increment the leading term of the polynomial, but we allow two new control parameters to enter the analysis, namely a_3 and a_4. This unfolding is called the 'butterfly catastrophe'. This is a five-dimensional surface whose intersection with (x, a_1, a_2)-space has three cusps that project onto the (a_1, a_2) as a curve that has the shape of a butterfly. Whence its name.

Chapter 7 The Butterfly Catastrophe

7.1 Definition of the Butterfly type

The butterfly type admits a rational unfolding:

$$E_{a_1,a_2,a_3,a_4} = \frac{1}{6}x^6 - \frac{1}{4}x^4 a_4 - \frac{1}{3}x^3 a_3 - \frac{1}{2}x^2 a_2 - x a_1 \tag{7.1.1}$$

and the associated catastrophe manifold M is given by the equations

$$\begin{aligned} 0 &= \frac{dE_{a_1,a_2,a_3,a_4}}{dx} \\ &= x^5 - x^3 a_4 - x^2 a_3 - x a_2 - a_1. \end{aligned} \tag{7.1.2}$$

This is a hypersurface in (a_1, a_2, a_3, a_4, x)-space. As figures in five-dimensional space can not be plotted, only projections of the hypersurface can be represented. The bifurcation set is obtained by differentiating (7.1.2) to obtain

$$5x^4 - 3x^2 a_4 - x a_3 = a_2 \tag{7.1.3}$$

and, substituting this back into (7.1.2), we get

$$-4x^5 + 2x^3 a_4 + x^2 a_3 = a_1. \tag{7.1.4}$$

For every pair of values (a_3, a_4) this gives a parameter equation of a bifurcation curve in the a_1, a_2 -plane. This curve is an isotangent curve (cf. section 2.2). One can get a picture of the full four-dimensional bifurcation set after K. Jänich, by looking at what happens above the curve $a_3^2 - a_4^2 = 1$ in the (a_3, a_4)-plane.

7.2 The Jänich representation

In figure 7.1, the bifurcation sets are shown for $a_3^2 - a_4^2 = 1$ in the (a_1, a_2)-plane with a_1 the horizontal and a_2 the vertical axis. For instance, when $a_3 = 0$ and $a_4 = -1$ then the bifurcation curve is cuspidal and may be considered as a (5,4)-cusp with control parameters a_1 and a_2.

Varying a_3 has the effect of swinging (or biasing) the whole picture sideways, the direction of the swing depending on the sign of a_3. Varying a_4 has the effect of adding a new cusp when a_4 becomes positive. For instance, the bifurcation curve shows a shape reminding of a butterfly when $a_3 = 0$ and a_4 is positive. Varying a_3 causes a similar bias distorting the butterfly.

To give an idea as to how the local extrema appear and disappear, we have drawn Figure 7.2 by setting the control parameters a_3, a_4 to the values 0 and 1, respectively. A two dimensional section of the energy function E_{a_1,a_2,a_3,a_4} as a function of x for some representative control points on the bifurcation curve are shown.

The bifurcation set separates the regions with a different number of extrema. Since the derivative of E is a polynomial of fifth degree, the maximal number of the real extrema is five. This number is reached when (a_1, a_2) lies in the so called *pocket*. Since the leading term is positive, two maxima and three minima appear. This means that in this region there are three stable and two unstable equilibria. Suppose that $a_3 = 0$,

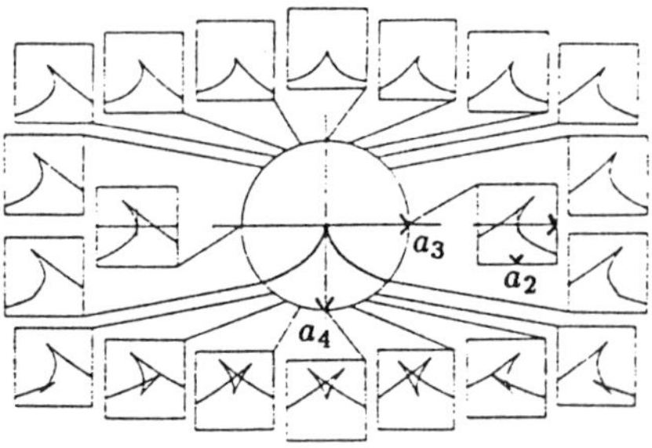

Fig. 7.1. The butterfly after K. Jänich (in Poston and Stewart(1978)

Fig. 7.2. E_{a_1,a_2,a_3,a_4} as function of x for some representative control points. (from Poston and Stewart(1978))

$a_4 = 1$ and that (a_1, a_2) is located just below the origin (e.g. (0,−0.05)). When we increase the value of a_1 to $a_1 = -0.1$ and keep $a_2 = -0.05$ constant, then the number of local extrema falls to three.

In this case we see one stable and one unstable equilibrium. If a_1 is again increased, then only one minimum remains; and we see one stable equilibrium. The picture is similar upon letting a_1 decrease. Figure 7.3 shows a two-dimensional section of M in the (a_1, x)-plane. It is a curve whose graph has equation

$$x^5 - x^3 + 0.05x = a_1. \tag{7.2.1}$$

For a_3 is equal to zero, this graph is point symmetrical with respect to the origin, and it folds four times making it consist of five sheets.

Projecting the graph onto the a_1-axis one distinguishes five regions on this axis. Above the middle region the graph has five sheets, whereas above the two neighbouring regions there are three sheets and in both directions farther away there is only one sheet. The number of sheets is equal to the number of local extrema of E. Around the origin one can imagine two types of orbits that a cursor may follow. The first one is analogous to the closed orbits of the previous chapters; it may start somewhere on the upper sheet, then slide to the left until it reaches the leftmost fold. Then by means of a catastrophic jump the cursor arrives at the lowest sheet, and follows this one until it reaches the rightmost fold. It then jumps to the upper sheet closing the orbit. It is clear that, as long as the control parameters a_2, a_3 and a_4 are kept constant then the cursor does not reach the middle sheet. The loop may be taken as being symmetrical with respect to the origin in the $(a_1, x) - plane$.

One may also start on the middle sheet and have the cursor oscillate around the origin. In that case, if the system is a damping system, the cursor will tend to the origin. If on the other contrary, the cursor moves away from the origin, and happens to reach the right fold it may jump to the extreme sheet and keep oscillating around the origin in the way previously described if the other control parameters are kept constant.

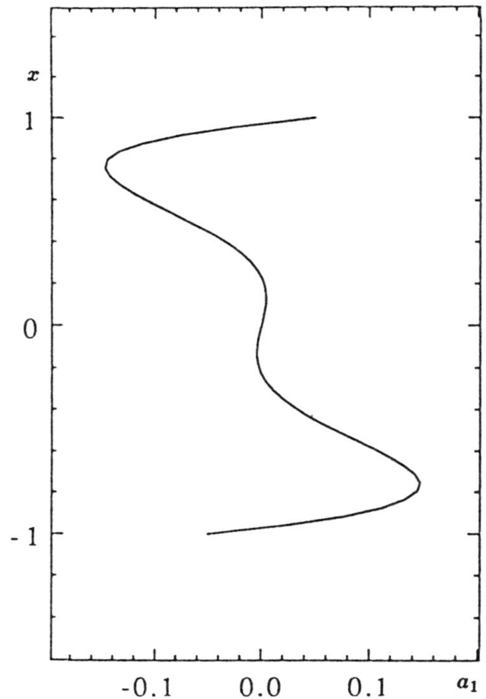

Fig. 7.3. Section of M in (a_1, x)-space for $a_2 = -0.05$, $a_3 = 0$ and $a_4 = 1$. (from [Poston and Stewart(1978)])

Fig. 7.4. Section of M in (a_1, a_2, x)-space for $a_3 = 0$ and $a_4 = 1$. (from [Poston and Stewart(1978)])

7.3 The case $a_3 = 0$ and $a_4 = 1$

If a_2 is allowed to vary, the orbit of the cursor may extend into ((a_1, a_2, x))-space on a surface. For instance, keeping the other control parameters fixed at $a_3 = 0$ and $a_4 = 1$ we obtain a surface which is shown in Figure 7.4. A characteristic section of this surface with the plane $a_2 = -0.05$ has already been shown in Figure 7.3; some other characteristic sections will be shown and discussed below.

Thus, a next section is taken for $a_2 = -0.22$, and is shown Figure 7.5. The curve also has four folds, and above $a_1 = 0$ there lie five sheets.

The equations analogous to (6.1.1-2), may be written as

$$\dot{a}_1 = \frac{-1}{K}x, \tag{7.3.1}$$

and

$$\dot{x} = -K(x^5 - x^3 + 0.22x - a_1). \tag{7.3.2}$$

The origin is a fixpoint for this system. If a_1 and x happen to be both zero, then also $\dot{a}_1$ and $\dot{x}$ are zero, meaning that the system may not escape from this state and the loop will never be closed.

Now, no closed loops are possible. Imagine a loop starting at the upper sheet. Since x is positive the cursor moves down and finally jump down to the middle sheet. The value of x is negative so that the cursor moves to the origin, which is a fixpoint. Starting from the lower sheet, for the same reason no closed loops are possible. Starting on the middle sheet, the cursor moves immediately to the origin.

With this set of differential equations and for constant values of the parameters $a_2 = -0.22$, $a_3 = 0$ and $a_4 = 1$ no closed loops are possible.

Next, let us consider a section at $a_2 = -0.3$. The corresponding graph is shown in Figure 7.7. Again no closed loops are possible, since the cursor is stopped dead at the origin on the middle sheet.

If on the other hand a_2 is smaller than -.45 there is only one sheet, and there are no catastrophic jumps at all. This is exemplified by Figure 7.7 where $a_2 = -1$.

Taking a_2 positive (e.g. $a_2 = 1$) the intersection curves resemble the graphs of the cusp model studied in the preceding chapters. These graphs are symmetric. Only a few situations corresponding to specific values of a_2 remain to be discussed. These values are $0, -.2, -.25$ and $-.45$. For a_2=0 we obtain the (5,4)-cusp which has the same shape as the the cusp-model of the previous chapter, except that there is a vertical tangent line in the origin. This graph is shown in Figure 7.9. In the case $a_2 = -.45$ the curve has no folds but only lines that are tangent to the x-axis. Hence, there are also no catastrophic jumps possible.

For $a_2 = -.2$ the solidly plotted curve (—) in Figure 7.8 obeys the equation

$$x^5 - x^3 + 0.2x = a_1. \tag{7.3.3}$$

It has four folds, but for this specific value of a_2 there are two pairs of folds, which lie just above each other. In case the cursor falls off the upper sheet, then there is 50%

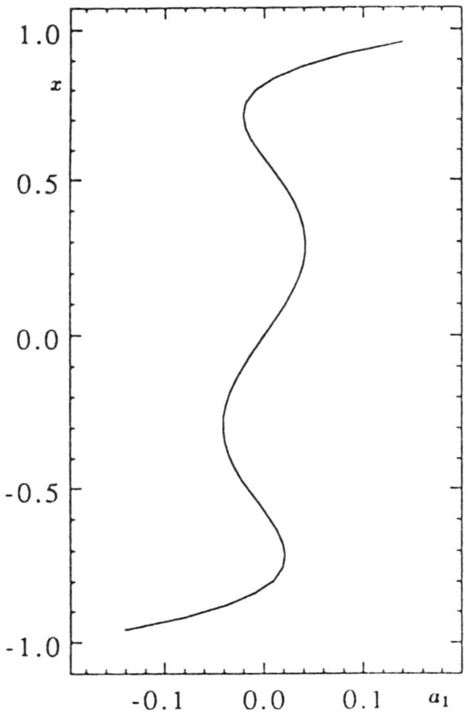

Fig. 7.5. M viewed in (a_1, x)-space at $a_2 = -0.22$, $a_3 = 0$ and $a_4 = 1$.

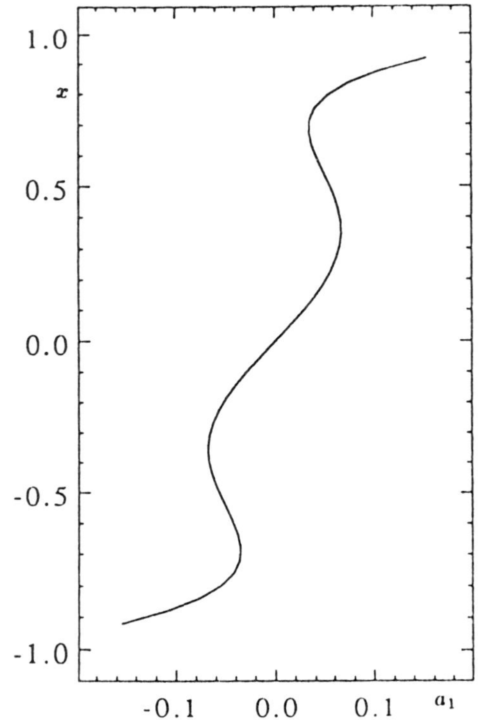

Fig. 7.6. M viewed in (a_1, x)-space at $a_2 = -0.3$, $a_3 = 0$ and $a_4 = 1$.

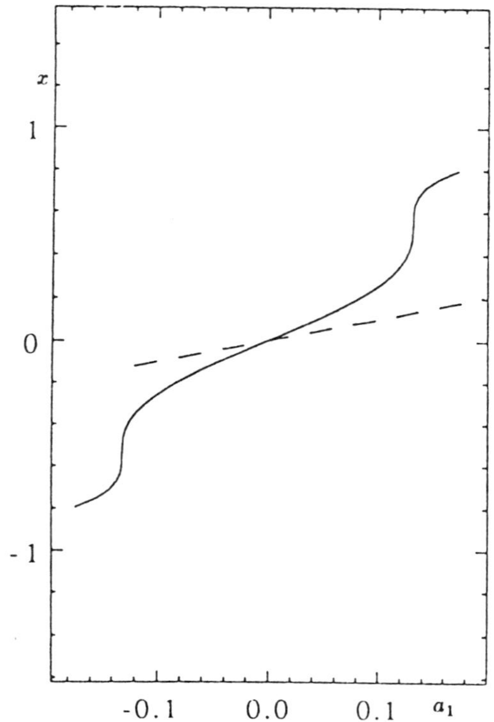

Fig. 7.7. M viewed in (a_1, x)-space for $a_2 = -1$ (- - -) and $-.45$(—), $a_3 = 0$ and $a_4 = 1$.

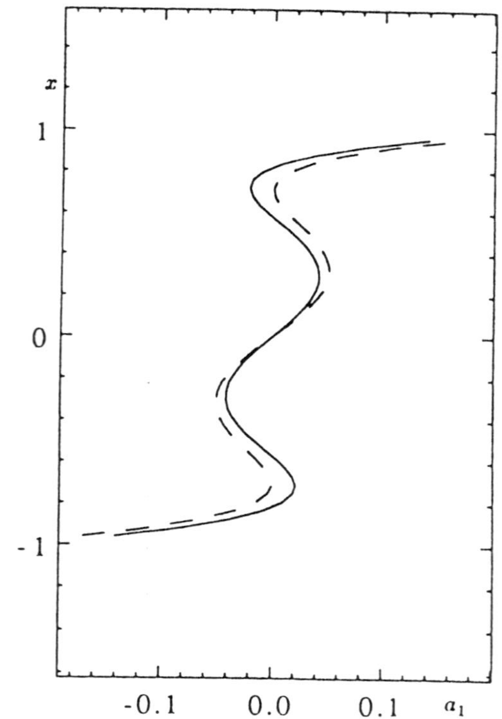

Fig. 7.8. M viewed in (a_1, x)-space for $a_2 = -0.2$(—)and $-.25$(- - -) $a_3 = 0$ and $a_4 = 1$.

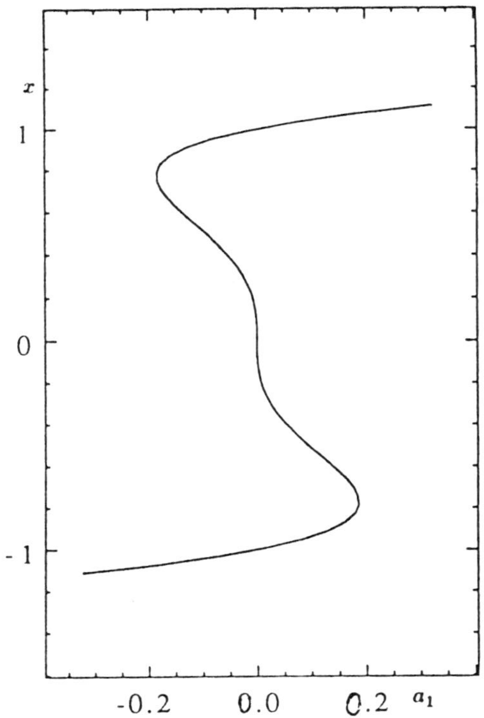

Fig. 7.9. M viewed in (a_1, x)-space for $a_2 = 0$, $a_3 = 0$ and $a_4 = 1$.

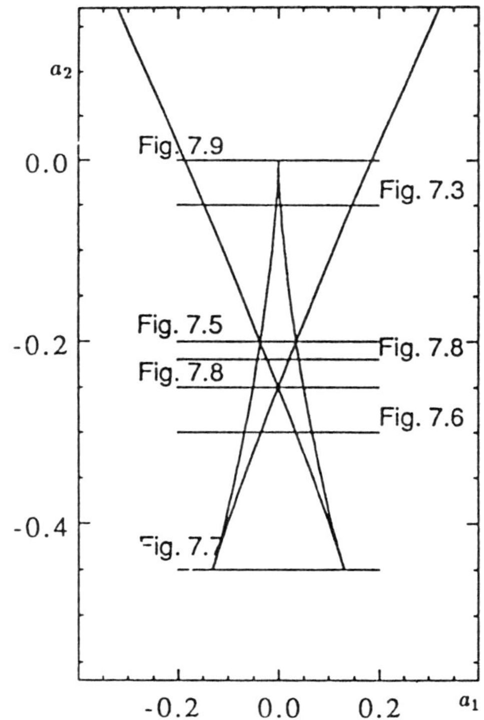

Fig. 7.10. Bifurcation set for $a_3 = 0$ and $a_4 = 1$ in the (a_1, a_2)-plane.

chance that it is stopped by the middle sheet and 50% that it travels to the lowest sheet. The orbit of the cursor is unpredictable. Finally, if $a_2 = -.25$ (- - -) than there are two folds lying precisely above the origin. If the cursor jumps from one of the extremal sheets to the middle sheet, then it will be trapped in the origin, and no closed loop is possible.

7.4 The general case $a_3 = 0$

When the control parameter $a_3 = 0$ then equation (7.1.2) becomes

$$a_1 = x^5 - x^3 a_4 - x a_2. \tag{7.4.1}$$

If the other control parameters (a_4 and a_2) are kept constant, equation (7.4.1) represents a curve in the (a_1, x)-plane which is symmetric with respect to the origin, since in the right-hand side there are only odd powers of x. The intersection points with the a_1-axis can be written down explicitly, and are found by equating the right-hand side of equation (7.1.2) to zero:

$$x = \sqrt{\frac{a_4 + \sqrt{D}}{2}} \quad \vee \quad x = \sqrt{\frac{a_4 - \sqrt{D}}{2}} \quad \vee$$
$$x = -\sqrt{\frac{a_4 + \sqrt{D}}{2}} \quad \vee \quad x = -\sqrt{\frac{a_4 - \sqrt{D}}{2}} \quad \vee$$
$$x = 0 \tag{7.4.2}$$

with

$$D = a_4^2 + 4a_2. \tag{7.4.3}$$

The number of intersection points depends on the value of a_4 and a_2. If $D < 0$ then there is only one real intersection point, namely the origin of the (a_1, x)-plane. If $D > 0$ then there are three cases. If $a_4 + \sqrt{D} < 0$ then again the curve intersects the a_1-axis only in the origin. If $a_4 + \sqrt{D} > 0$ and $a_4 - \sqrt{D} < 0$ then there are three intersection points. If one of $a_4 - \sqrt{D} > 0$ then there are five intersection points. These crossing points lie pairwise (point) symmetrical with respect to the origin. The maximal distance d_z between the roots is given by

$$d_z = 2\sqrt{\frac{a_4 + \sqrt{D}}{2}} \qquad \text{if} \qquad a_4 + \sqrt{D} > 0 \tag{7.4.4a}$$

$$\text{else} \qquad d_z = 0. \tag{7.4.4b}$$

The distance (d_z) between the uppermost and lowest sheet along the x-axis is increasing as a function of a_4 and a_2. If $a_2 > 0$ then $d_z > 0$, and

$$\lim_{a_4 -> -\infty} d_z = 0.$$

The jump points lie on the bifurcation set of the manifold and are given by :

$$5x^4 - 3x^2 a_4 = a_2 \tag{7.4.5}$$

and

$$-4x^5 + 2x^3 a_4 = a_1. \tag{7.4.6}$$

If a_4 is smaller than or equal to zero then the bifurcation set is a (5,4)-cusp, whereas if a_4 is larger than zero then the bifurcation set has the actual butterfly shape. The x-coordinates of the jump points can be written explicitly as follows:

$$x = \sqrt{\frac{3a_4 + \sqrt{D_3}}{10}} \quad \vee \quad x = \sqrt{\frac{3a_4 - \sqrt{D_3}}{10}} \quad \vee$$

$$x = -\sqrt{\frac{3a_4 + \sqrt{D_3}}{10}} \quad \vee \quad x = -\sqrt{\frac{3a_4 - \sqrt{D_3}}{10}} \tag{7.4.7}$$

with

$$D_3 = 9a_4^2 + 20a_2. \tag{7.4.8}$$

If $D_3 < 0$ or ($D_3 > 0$ and $3a_4 + \sqrt{D_3} < 0$) then there are no jump points. If $D_3 > 0$ and $3a_4 + \sqrt{D_3} > 0$ then there are at least two jump points. In the latter case, the maximal distance between the jump points (d_{jx}) in the x-direction is given by:

$$d_{jx} = 2\sqrt{\frac{3a_4 + \sqrt{D_3}}{10}}$$

$$\text{else} \qquad d_{jx} = 0, \tag{7.4.9}$$

which is also an increasing function of a_4 and a_2.

7.5 The case $a_3 \neq 0$

If the control parameter a_3 is different from zero, then the graph in the (a_1, x)-plane is no longer symmetrical, and the intersection points of the x-axis with the curve are given by

$$x = \frac{a_3}{4\sqrt{t}} - \sqrt{\frac{{a_3}^2}{4t} + \sqrt{t} - \frac{s}{2}} \quad \vee \tag{7.5.1a}$$

$$x = \frac{a_3}{4\sqrt{t}} + \sqrt{\frac{{a_3}^2}{16t} + \sqrt{t} - \frac{s}{2}} \quad \vee \tag{7.5.1b}$$

$$x = \frac{-a_3}{4\sqrt{t}} + \sqrt{\frac{{a_3}^2}{16t} - \sqrt{t} - \frac{s}{2}} \quad \vee \tag{7.5.1c}$$

$$x = \frac{-a_3}{4\sqrt{t}} - \sqrt{\frac{{a_3}^2}{16t} - \sqrt{t} - \frac{s}{2}} \quad \vee \tag{7.5.1d}$$

$$x = 0, \tag{7.5.1e}$$

with

$$u = \frac{-{a_4}^3}{27} + \frac{2a_4 a_2}{3} - \frac{-{a_3}^2 + 4a_4 a_2}{2} \tag{7.5.2}$$

$$v = \frac{-{a_4}^2}{9} + \frac{4a_2}{3} \tag{7.5.3}$$

$$w = \left(u + \sqrt{(v)^3 + (u)^2}\right)^{\frac{1}{3}} \tag{7.5.4}$$

$$s = \frac{-a_4}{3} - \frac{v}{w} + w \tag{7.5.5}$$

$$t = a_2 + \frac{(s)^2}{4}. \tag{7.5.6}$$

Note that if a_3 is replaced by $-a_3$ then equations 7.5.1e-6 remain the same while the roots 7.5.1a,b interchange with 7.5.1c,d. Thus the sign of a_3 does not influence the intersection points with the x-axis. We have shown in the previous paragraph that the maximal distance between the curve and the intersection points with the x-axis is an increasing function of a_4. This is also true if $a_3 \neq 0$. Define a function $f(x)$ by:

$$f(x, a_4) = x^5 - x^3 a_4 - x^2 a_3 - x a_2. \tag{7.5.7}$$

The x-coordinates are the roots of the equation

$$f(x, a_4) = 0. \tag{7.5.8}$$

If $(0, a_2, a_3, a_4)$ does not lie on the bifurcation set of the manifold, then the roots of a polynomial are continuous functions of its coefficients. Thus, if a_4 is changed to $(a_4 + \Delta a_4)$, so that $(0, a_2, a_3, a_4)$ is not separated from $(0, a_2, a_3, a_4 + \Delta a_4)$ by the bifurcation set, then a root x_0 of (7.5.8) changes continuously to $(x_0 + \Delta x_0)$ and we have:

$$f(x_0 + \Delta x_0, a_4 + \Delta a_4) = 0. \tag{7.5.9}$$

This gives

$$\begin{aligned} &(x_0 + \Delta x_0)^5 - (x_0 + \Delta x_0)^3 (a_4 + \Delta a_4) - \\ &(x_0 + \Delta x_0)^2 a_3 - (x_0 + \Delta x_0) a_2 = 0 \end{aligned} \tag{7.5.10}$$

and

$$x_0^5 - x_0^3 a_4 - x_0^2 a_3 - x_0 a_2 = 0. \tag{7.5.11}$$

Subtracting eq. 7.5.11 from eq. 7.5.10 and neglecting the term in $\Delta a_4 \Delta x_o$ and higher order terms in Δa_4 and Δx_0, we obtain

$$\Delta x_0 f'_x(x_0, a_4) = \Delta a_4 x_0^3. \tag{7.5.12),}$$

where $f'_x(x_0, a_4)$ is the partial derivative of f with respect to x. As the curve passes through the origin of the (a_1, x)-plane, there is at least one real root of eq. 7.5.11. We define the largest and smallest real root as

$$x_+ = \max\{x_0 | f(x_0, a_4) = 0, x_0 \text{ real}\} \tag{7.5.13}$$

$$x_- = \min\{x_0 | f(x_0, a_4) = 0, x_0 \text{ real}\}. \tag{7.5.14}$$

Since eq.7.5.7 is of the order five, and the coefficient of the highest term is positive, the derivatives at (7.5.13) and (7.5.14) are positive. This means that $f'_x(x_+, a_4) > 0$ and $f'_x(x_-, a_4) > 0$. The fact that $x_0 = 0$ is a root of equation 7.5.11 means $x_+ \geq 0$ and

$x_- \leq 0$. Hence, the difference in distance d_z between x_+ and x_- changes with an amount Δd_z.

If $x_+ > 0 > x_-$ then Δd_z is given by:

$$\begin{aligned} \Delta d_z &= \Delta x_+ - \Delta x_- \\ &= \left(\frac{x_+^3}{f'_x(x_+, a_4)} - \frac{x_-^3}{f'_x(x_-, a_4)}\right)\Delta a_4. \end{aligned} \tag{7.5.15}$$

Since the coefficient of Δa_4 is positive, the $\Delta d_z = \Delta x_+ - \Delta x_-$ is proportional to Δa_4.

If $x_+ = 0 = x_-$ then $d_z = 0$ and is independent of a_4.

If $x_+ = 0$ and $x_- < 0$ then $\Delta d_z = -\Delta x_-$ which is proportional to Δa_4.

If $x_- = 0$ and $x_+ > 0$ then $\Delta d_z = \Delta x_+$ which is proportional to Δa_4.

This means that if the value of a_4 increases then the height between the lowest and highest intersection points of the x-axis with the manifold increases or at least remains the same. If the value of a_4 decreases then this height decreases or at least remains the same. The previous result is more general and can be formulated to hold for all values of n.

7.6 Distance between extreme roots of a polynomial equation as a function of the coefficients

We study polynomials of the form:

$$x^{n+1} - a_n x^{n-1} - \ldots - a_2 x = a_1. \tag{7.6.1}$$

The term in x^n is missing and we are interested in the distance between the extremal intersection points of the polynomial curve with the x-axis, everything being measured in the (a_1, x)-plane. Define a function $g(x)$ by:

$$g(x) = x^{n+1} + a_n x^{n-1} + \ldots + a_2 x, \tag{7.6.2}$$

and rewrite (7.6.1) as follows:

$$g(x) = a_1. \tag{7.6.3}$$

The curve goes through the origin (a_1, x)-plane, all the intersection points with the x-axis are given by:

$$g(x) = 0. \tag{7.6.4}$$

As in the previous paragraph we suppose that x_0 is a root of equation 7.6.4 and that $g'_x(x_0) \neq 0$. The root x_0 is a continuous function of the coefficients $a_2, \ldots, a_n$. If a_n changes while the other coefficients are kept constant, then the root x_0 will change as well. In case $(0, a_2, \ldots, a_n)$ is not separated from $(0, a_2, \ldots, a_n + \Delta a_n)$ by the bifurcation set, for a change Δa_n, the root x_0 will change to $(x_0 + \Delta x_0)$. The value of Δx_0 can be approximated by replacing x_0 by $(x_0 + \Delta x_0)$ and a_n by $(a_n + \Delta a_n)$ in equation 7.6.1. We get :

$$(x_0 + \Delta x_0)^{(n+1)} - (x_0 + \Delta x_0)^{(n-1)}(a_n + \Delta a_n) - \ldots$$
$$-(x_0 + \Delta x_0)a_2 = 0. \quad (7.6.5)$$

Subtracting the original equation 7.6.2 from 7.6.5 and neglecting $\Delta a_n \Delta x_o$ and higher order terms in Δa_n, and Δx_0 we obtain

$$[(n+1){x_0}^{(n-1)} - (n-1){x_0}^{(n-2)}a_n - \ldots - a_2]\Delta x_0 -$$
$$\Delta a_n {x_0}^{(n-1)} = 0, \quad (7.6.6)$$

which is equal to

$$g'_x(x_0)\Delta x_0 - \Delta a_n {x_0}^{(n-1)} = 0. \quad (7.6.7)$$

Define the largest and smallest ("extremal") real root as

$$x_+ = \max\{x_0 | g(x_0) = 0, x_0 \text{ real}\}, \quad (7.6.8)$$

$$x_- = \min\{x_0 | g(x_0) = 0, x_0 \text{ real}\}. \quad (7.6.9)$$

We distinguish two cases depending on the order n of eq. 7.6.2.

If n is *odd*, since the coefficient of the highest order term is positive, then $g'_x(x_+) > 0$ and $g'_x(x_-) < 0$. The root $x_0 = 0$ is independent of the value of a_n, thus $x_+ \geq 0$ and $x_- \leq 0$. The difference in distance d_z between the largest root x_+ and smallest x_- changes with an amount Δd_z. Depending on the value of x_+ and x_- we distinguish four cases:

If $x_+ > 0$ and $x_- < 0$ then $x_-^{(n-1)} > 0$ and

$$\Delta d_z = \Delta x_+ - \Delta x_-$$
$$= \left(\frac{x_+^{(n-1)}}{g'_x(x_+)} - \frac{x_-^{(n-1)}}{g'_x(x_-)}\right)\Delta a_n. \quad (7.6.10)$$

If $x_+ = 0$ and $x_- < 0$ than $x_-^{(n-1)} > 0$ and $g'_x(x_-) < 0$,thus

$$\Delta d_z = -\Delta x_- = \left(-\frac{x_-^{(n-1)}}{g'_x(x_-)}\right)\Delta a_n \quad (7.6.11)$$

has the same sign as Δa_n.

If $x_- = 0$ and $x_+ > 0$ than $g'_x(x_+) > 0$,thus

$$\Delta d_z = \Delta x_+ = \left(\frac{x_+^{(n-1)}}{g'_x(x_+)}\right)\Delta a_n \quad (7.6.12)$$

has the same sign as Δa_n.

If $x_- = 0$ and $x_+ = 0$ than

$$\Delta d_z = 0, \quad (7.6.13)$$

and is independent of the value of a_n.

If at least one of the extremal roots is different from zero, the sign of the difference between the extremal roots is the same as the sign of the change in a_n. The distance remains zero otherwise.

If n is *even* then $g'_x(x_+) > 0$ and $g'_x(x_-) > 0$. Again, there are four possibilities for the value of the extreme roots x_+ and x_- , and we distinguish four cases:

If $x_+ > 0$ and $x_- < 0$ then $x_-^{(n-1)} < 0$ and

$$\Delta d_z = \Delta x_+ - \Delta x_- = (\frac{x_+^{(n-1)}}{g'_x(x_+)} - \frac{x_-^{(n-1)}}{g'_x(x_-)})\Delta a_n. \tag{7.6.14}$$

If $x_+ = 0$ and $x_- < 0$ then $x_-^{(n-1)} < 0$, thus

$$\Delta d_z = -\Delta x_- = (-\frac{x_-^{(n-1)}}{g'_x(x_-)})\Delta a_n \tag{7.6.15}$$

has the same sign as Δa_n.

If $x_- = 0$ and $x_+ < 0$, then

$$\Delta d_z = \Delta x_+ = (\frac{x_+^{(n-1)}}{g'_x(x_+)})\Delta a_n \tag{7.6.16}$$

has the same sign as Δa_n.

If $x_- = 0$ and $x_+ = 0$ then

$$\Delta d_z = 0, \tag{7.6.17}$$

and is independent of the value of a_n.

We may conclude for every order of the polynomial of equation 7.6.1 : If at least one of the extremal roots is different from zero, then the distance between the extremal roots changes in the same sense as a_n. The distance remains zero otherwise.

Due to the shape of the formulas 7.6.1-17 this result holds also for $a_{(n-2)}, \ldots, a_{(n-2k)}$ with $(n-2k > 0)$. This means that for $n = 4$ the distance between the intersection points of the x-axis and the manifold M becomes larger when the control parameters a_4 and a_2 are both increasing. The distance decreases when a_4 and a_2 are both decreasing. When one of the parameters is increasing while the other is decreasing, the effects are conflicting, which results in a partial increase or decrease.

This means that both parameters are interacting and allow a sensitivity tuning, meaning that both control parameters are linked.

If a_n changes in $a_{(n-1)}$ then eq. (7.6.7) still holds. As the curve goes through the origin $x_+ \geq 0$ and $x_- \leq 0$. Depending on the order n of eq. (7.4.2) we distinguish an odd and an even case.

If n is *odd*, since the coefficient of $x^{(n+1)}$ is positive, we have $g'_x(x_+) > 0$ and $g'_x(x_-) < 0$, and $x_-^{(n-2)} \leq 0$ and $x_+^{(n-2)} \geq 0$, thus x_- and x_+ remain constant or change in the same sense as $a_{(n-1)}$.

If n is *even*, since the coefficient of $x^{(n+1)}$ is positive we have $g'_x(x_+) > 0$ and $g'_x(x_-) > 0$, and $x_-^{(n-2)} \leq 0$ and $x_+^{(n-2)} \geq 0$, thus x_- and x_+ remain constant or change in the same sense as $a_{(n-1)}$.

Summing up: x_- and x_+ remain constant or change in the same sense as $a_{(n-1)}$, and this holds also for every $a_{(n-2k-1)}$ where k is any integer satisfying $k \geq 0$ and $n - 2k - 1 > 1$.

We conclude that :

x_+ remains constant or changes in the same sense as a_k for every integer k with $n \geq k > 1$.

x_- remains constant or changes in the same sense as a_{n-2k-1} for every integer $k \geq 0$ and $n - 2k - 1 > 1$.

x_- remains constant or changes in the opposite sense as a_{n-2k} for every integer $k \geq 0$ and $n - 2k > 1$.

7.7 The (5,4)-cusp ($a_3 = a_4 = 0$)

When the control parameters a_3 and a_4 are both zero, a (5,4)-cusp is obtained with the equation:

$$a_1 = x^5 - xa_2. \tag{7.7.1}$$

If the control parameter a_2 is kept constant, equation (7.7.1) represents a curve in the (a_1, x)-plane which is symmetric with respect to the origin. If a_2 is positive, the real intersection points with the x-axis can be written down explicitly, by setting a_4 to zero in equation (7.4.2) :

$$x = \sqrt[4]{a_2} \quad \vee \quad x = 0 \quad \vee \quad x = -\sqrt[4]{a_2} \tag{7.7.2}$$

The jump points lie on the bifurcation set of the manifold and are given by :

$$a_2 = 5x^4 \tag{7.7.3}$$

and

$$a_1 = -4x^5. \tag{7.7.4}$$

The x-coordinates of the jump points can be written explicitly as follows:

$$x_j = \sqrt[4]{a_2/5} \quad \vee \quad x_j = -\sqrt[4]{a_2/5} \tag{7.7.5}$$

There are two jump points. The corresponding a_1-value for the positive x_j is:

$$\alpha_1 = -(4/5)a_2\sqrt[4]{a_2/5}. \tag{7.7.6}$$

The distance between the jump point and the lowest sheet in the x-direction can be found by equating a_1 to α_1 in eq. (7.7.1) and calculating the roots of this equation. This equation of fifth degree in x can be solved because the positive x_j is a double root of this equation. Thus the equation can be reduced to a third order one, which has one real root if a_2 is positive, which is given by

$$x_e \approx -1.10384\sqrt[4]{a_2}. \tag{7.7.7}$$

The distance in the x-direction between the jump point and the opposite sheet is given by

$$1.10384\sqrt[4]{a_2} + \sqrt[4]{a_2/5}. \tag{7.7.8}$$

Previous findings will be applied to the left ventricle, in the next chapters.

Chapter 8 5-D Catastrophe Model for the Ventricle

8.1 General remarks

In Chapter 6 we elaborated a first-order catastrophe model of the cardio-vascular system. The model was of the cusp type and we assigned the state parameter x to the ventricular pressure plus a constant. The constant was chosen so that the minimal pressure lies on the resting volume-pressure curve.

The control parameters a_1 and a_2 were identified with a dynamic parameter and the ventricular volume (v) respectively.

As explained in Chapter 1, the cardio-vascular system is at least controlled by the heart rate, the preload, the myocardial contractility and the afterload. The first two parameters were already included in the cusp-model.

The rate is determined by the first control parameter a_1, which forces the catastrophic jumps. The time derivative of this parameter $\dot{a}_1$ was equal to $-Sx/K$, where S is a time scaling parameter, and K a constant related to the catastrophic jumps. This parameter determines the duration of the systole and diastole. In these cusp models, x is a symmetric function, meaning that the duration of the diastole is equal to that of the systole. This is a good approximation in a resting man (as explained in Chapter 5), but as the heart rate increases, the duration of the diastole shortens more than that of the systole. A time difference between systole and diastole in the cusp models can be forced in the model if we allow two time scalings S_d and S_s during diastole and systole, respectively.

A second control parameter is the preload. The preload is quantified by the v. When the ventricular volume is large, the stretching of the heart muscle fibre is large. The "Frank-Starling Law" relates the stretching of the fibre to the performance of the heart. This feature has already been included in the cusp models.

The cusp models have only two control parameters, whereas the cardio-vascular system is controlled by at least four parameters. This suggests an extension of the model to a butterfly catastrophe model. The latter model allows two more parameters (a_3 and a_4).

In Chapter 7 we have shown that the parameters a_2 and a_4 are linked. If one of these parameters increases and the other stays constant, or when both increase, then the distance between the stable sheets increases. Thus the distance between x_{min} and x_{max} increases.

8.2 Setting a_3 equal to a constant different from zero

In the following graphs we show the butterfly catastrophe with the same control parameters a_1 and a_2 =v, as for the (5,4)-cusp model, but with a_3 equal to a constant different from zero (e.g. a_3=1). / Since a_3 is constant, the equation (6.3.2) still holds in this case, but is renumbered for convenience as

$$\dot{a}_1 = \frac{-S}{K}x. \tag{8.2.1}$$

The oscillator formula (6.3.1) takes the shape

$$\dot{x} = -SK(x^5 - x^2 - vx - a_1). \tag{8.2.2}$$

The values of S and K are kept equal to their values in the cusp situation. We calculated the graphs for $a_3 = 1$, and draw the (v,p)-, (a_1,x)-, (t,v)-, (v,f)- and (f,p)-plots. In all these plots the general behaviour is the same as in the (5,4)-cusp model (for a_3 is equal to zero). To show the minor differences between these two models we draw them on top of each other. The butterfly model is drawn in full lines while the (5,4)-cusp is drawn in broken lines.

The (v,p)-plot (Fig. 8.1) shows a broadening and a slight pressure raise when $a_3 = 1$. The EDV remains approximately the same but the end systolic volume ESV is decreased, so that the stroke volume (SV) increases. The peak systolic pressure (PSP) is attained when v is close to EDV in both cases.

The largest differences appear in the (a_1,x)-plot (Fig. 8.2). In the cusp model the graph is always symmetrical over the origin in this plane. But when a_3 is different from zero the graph is no longer symmetrical over the origin. As shown in the previous chapter the graph is swung to the right when the sign of a_3 is positive. The lower (diastole) and upper (systole) part of this graph are different, causing a change in timing.

The (v,f)-patterns (Fig. 8.3) are very similar. In the case of the cusp model, both volume and flow have slightly lower values in comparison to the butterfly model.

The (f,p)-graph (Fig. 8.4) has two levels. In the low level the flow is reduced only in the case $a_3 = 0$, while the pressure remains the same; this is due to a translation of the pressure letting the minimal pressure lie on the resting volume-pressure curve. The upper level shows an inclined part followed by a narrow loop. The errors on the inclined part are large, because of the fast jump of both, pressure and flow. The narrow loop is due to the slowly decreasing pressure during the rest of systole.

In the time-pressure plot (Fig. 8.5) we observe an increase of the period from 0.8 to 1.0 as the value of a_1 increases from 0 to 1. We denote the duration of the systole and diastole by T_s and T_d, respectively. We observe that in the case of the (5,4)-cusp (with $a_3 = 0$ and $a_4 = 0$) that T_s is equal to T_d. As a_3 increases from 0 to 1 and a_4 remains zero, then T_s remains the same while T_d increases. Thus the ratio T_s/T_d becomes smaller than one.

8.3 Setting a_4 equal to a constant different from zero

In this paragraph we comment on the simulations of a butterfly catastrophe with the same parameters a_1 and $a_2 = EDV$ as in Chapter 6, but with $a_3 = 0$ and $a_4 = 2$. The slow equation remains the same as eq. 8.2.2 but the fast equation changes in:

$$\dot{x} = -SK(x^5 - 2x^3 - EDVx - a_1). \tag{8.3.1}$$

The values of S and K are kept equal to their values in the cusp situation.

In Figure 8.6, we observe an increase in pressure during systole as well as in stroke volume. The pressure rise is due to the increasing distance between the stable sheets. The peak systolic pressure is still reached near EDV. The broadening of the graph is a consequence of the lowered systolic volume, causing a lower minimal pressure and an

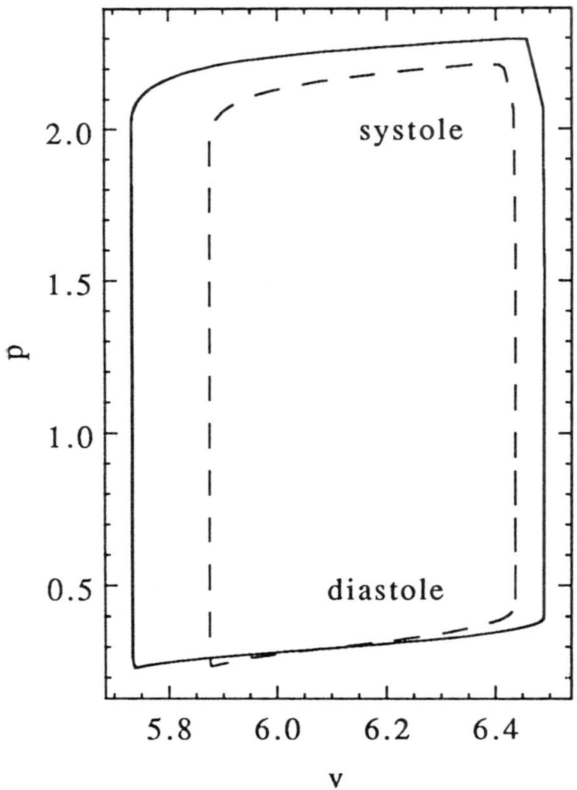

Fig. 8.1. (v, p)- graph for a butterfly catastrophe with ($a_2 = v$ and $a_3 = 1$(—) or $a_3 = 0$ (- - -)).

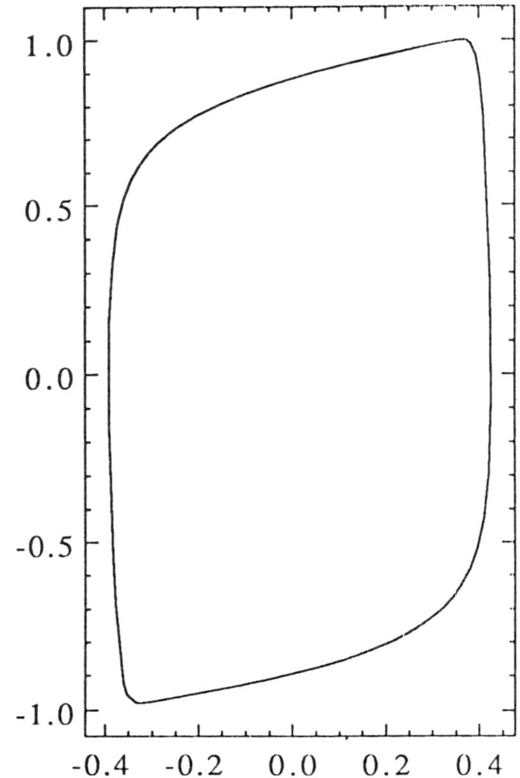

Fig. 8.2. (a_1, x)-plot for a butterfly catastrophe with ($a_2 = v$ and $a_3 = 1$(—) or $a_3 = 0$(- - -)).

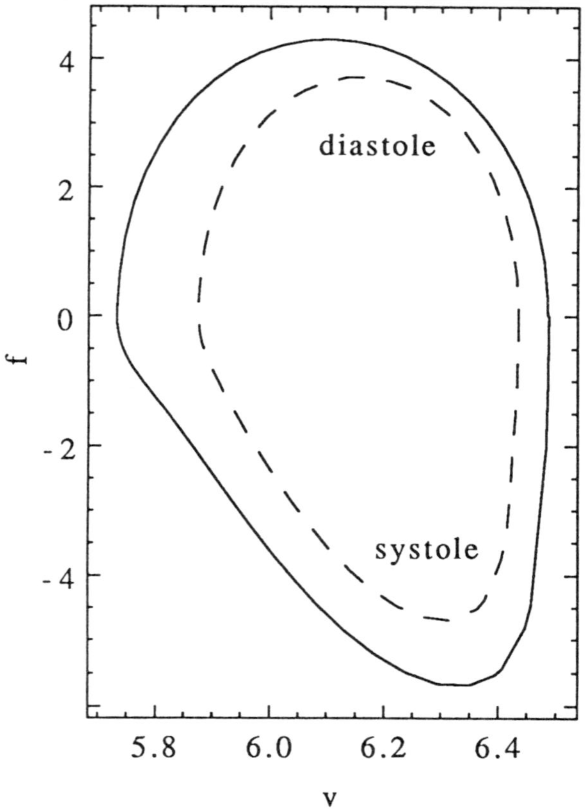

Fig. 8.3. (v, f)- graph for a butterfly catastrophe with ($a_2 = v$ and $a_3 = 1$(—) or $a_3 = 0$(- - -)).

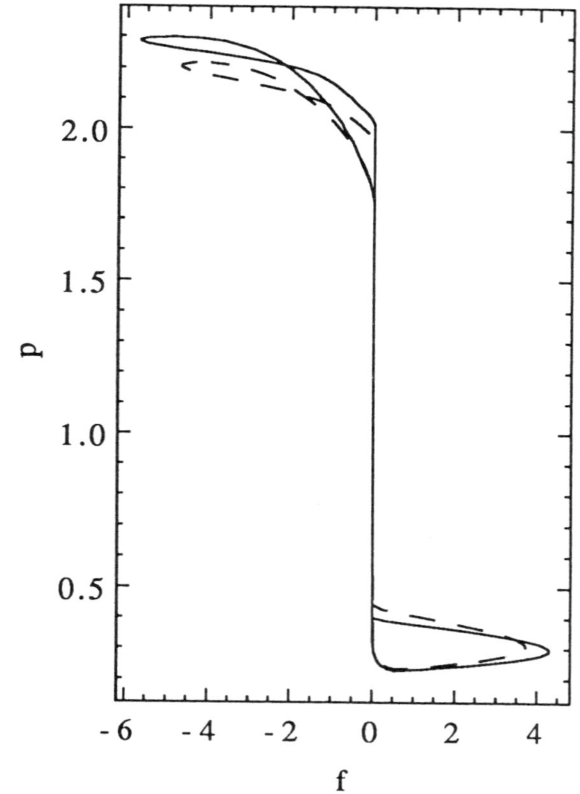

Fig. 8.4. (f, p)- graph for a butterfly catastrophe with ($a_2 = v$ and $a_3 = 1$(—) or $a_3 = 0$(- - -)).

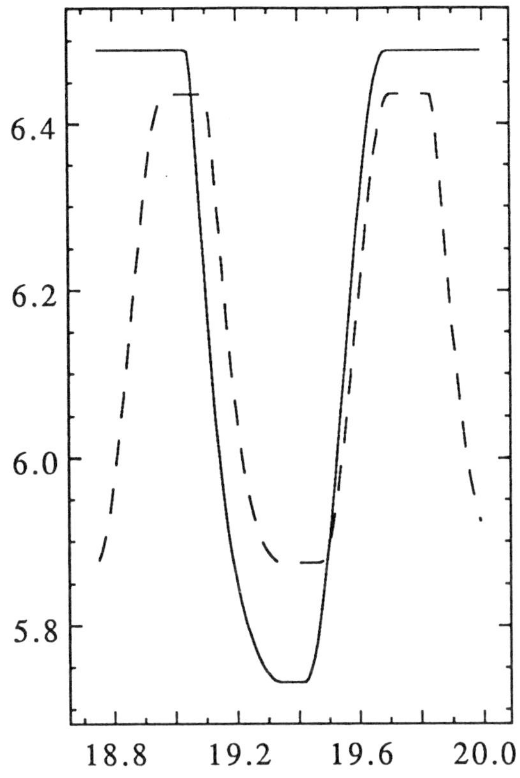

Fig. 8.5. (t, v)- graph for a butterfly catastrophe with ($a_2 = v$ and $a_3 = 1$(—) or $a_3 = 0$(- - -)).

inflated inflow during diastole. The increased stroke volume is also due to a substantial lowering of the beat rate.

Another remark is that the (a_1, x)-plot (Fig. 8.7) is symmetrical over its origin as in the cusp model; this is due to the absence of a term with an even power of x. This means that T_s is equal to T_d. Due to the value of a_4 this graph is larger than the (a_1, x)-plot (Fig. 8.2) of the cusp model. The catastrophic jumps occur at larger values of x and a_1, causing an increase in period.

This mismatch of timing is very well shown in the (v, f)-plot (Fig. 8.8). At the left end there appears a tail. The volume decreases slowly (small negative flow). This is due to an outflow through the arteries at the time that the pressure is almost equal to the arterial pressure.

The (f,p)-plot (Fig. 8.9) is very similar to the preceding (f,p)-plots (see e.g. Fig. 8.4). Only the pressure is raised in a substantial way.

The (t,p)-plot (Fig. 8.10) is a time graph. It shows again the time mismatch during systole. The volume decreases first rapidly and then slowly to reach ESV. The pressure wave during the diastole has the right shape but the period is far too large. Note that the diastole and systole have the same duration, as explained earlier.

8.4 The parameter a_4 as the afterload

To obtain a (v,p)-graph with a peak systolic pressure located in the middle between EDV and ESV, we identify the control parameter a_4 with the afterload. The afterload is quantified using the aortic pressure P_{aor}. The fast equation becomes

$$\dot{x} = -SK(x^5 - P_{\text{aor}}x^3 - vx - a_1), \qquad (8.4.1).$$

The slow equation remains the same:

$$\dot{a}_1 = \frac{-S}{K}x. \qquad (8.4.2)$$

Note that P_{aor} has the same period as x because they are related by a set of linear differential equations. The aortic pressure P_{aor} is oscillating during the first part of systole and decreases elsewhere. This introduces a difference in duration between the systole and diastole. In Figure 8.11 we show a volume-time plot for that situation. The values of the duration of the systole and diastole are $T_s = 1.4s$ and $T_d = 1.7s$, respectively. The ratio of T_s/T_d, is no longer 1 but becomes 0.8. The period is $T_s + T_d = 3.1s$. To obtain an acceptable period the value of S is adjusted from 14.5 to 75.

For $S = 75$ we made a (v,p)-plot, which is shown in Figure 8.12. This graph resembles very well the (v,p)-graph simulated with Rideout's model of the heart (Chapter 5 , Figure 5.12). In both simulations the fast pressure jumps at constant volume can be seen. At high pressure level, the graphs show a fast emptying, while the pressure is still increasing, followed by a slower volume decrease, while the pressure is decreasing. PSP is reached between EDV and ESV. Only the bulge at the end of the diastole in the recorded graph (cf. Fig. 5.13) is not obtained. This is due to the fact that the atrial contraction has not been included in our simulation model.

The (a_1,x)- graph (Fig 8.13) is not symmetrical any more, although the characteristic equation of the catastrophe model contains a_1 and only odd terms in x. The asymmetry is caused by the fact that the control parameter a_4 changes during one cycle. In the previous models only the control parameter a_1 and the state parameter x changed during the cycle.

The simulated (v,f)-plot (Fig. 8.14) shows the same graph as the simulated (v,f)-plot with Rideout's model (Fig. 5.14). From EDV the flow falls to a large negative value. Then the flow decreases to zero while the aortic valve closes. When the AV-valve opens the inflow into the ventricle increases and decreases again when the ventricular pressure exceeds the venous pressure. The simulated graph does not show the supplementary structure (Fig. 5.15) which is due to atrial contraction.

The simulated (f,p)-graph after the catastrophe model (Fig. 8.15) resembles very well the (f,p)-graph after Rideout's model (Fig. 5.16). Both show differences with the

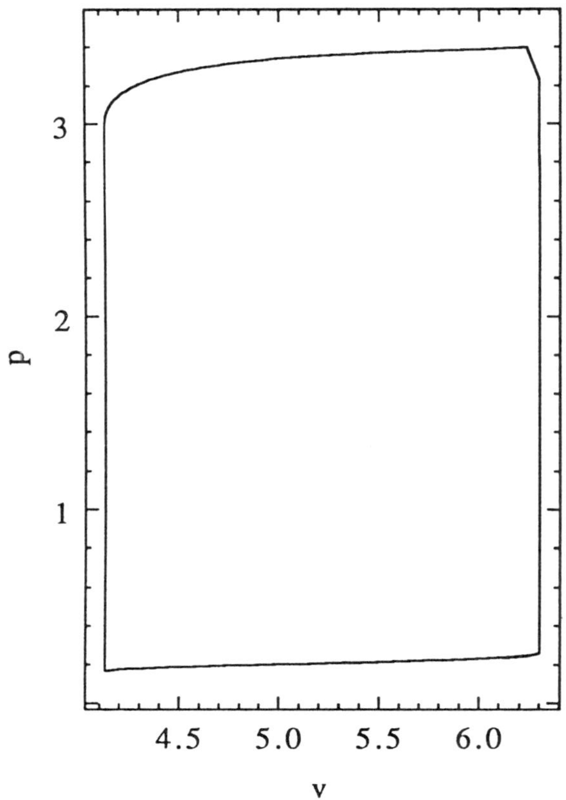

Fig. 8.6. (v, p)- graph for a butterfly catastrophe with ($a_2 = v$ and $a_4 = 2$)

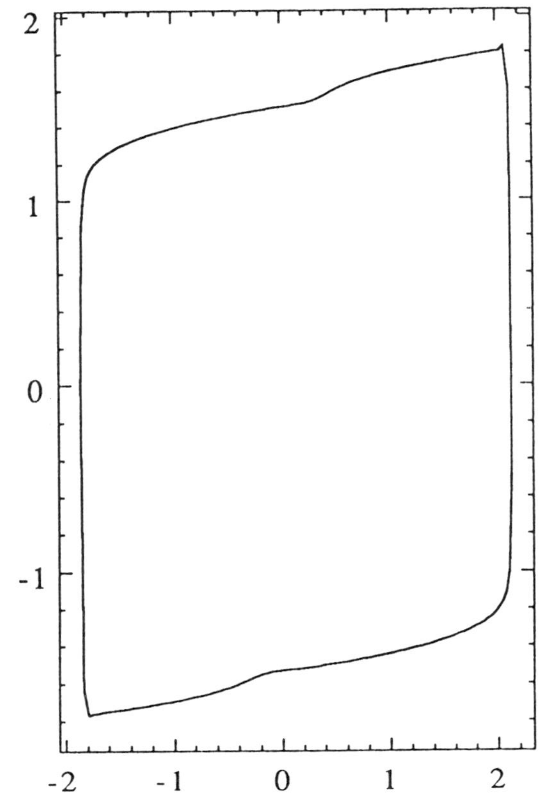

Fig. 8.7. (a_1, x)-plot for a butterfly catastrophe with ($a_2 = v$ and $a_4 = 2$)

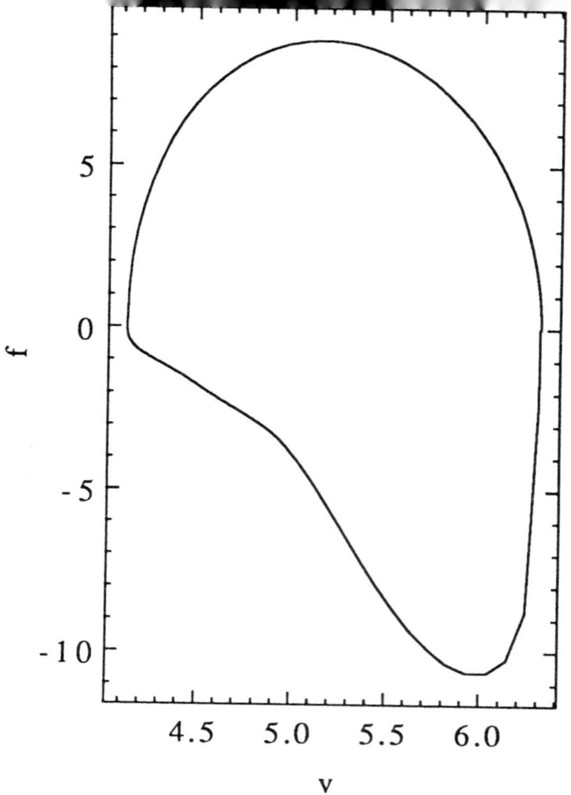

Fig. 8.8. (v, f)- graph for a butterfly catastrophe with ($a_2 = v$ and $a_4 = 2$)

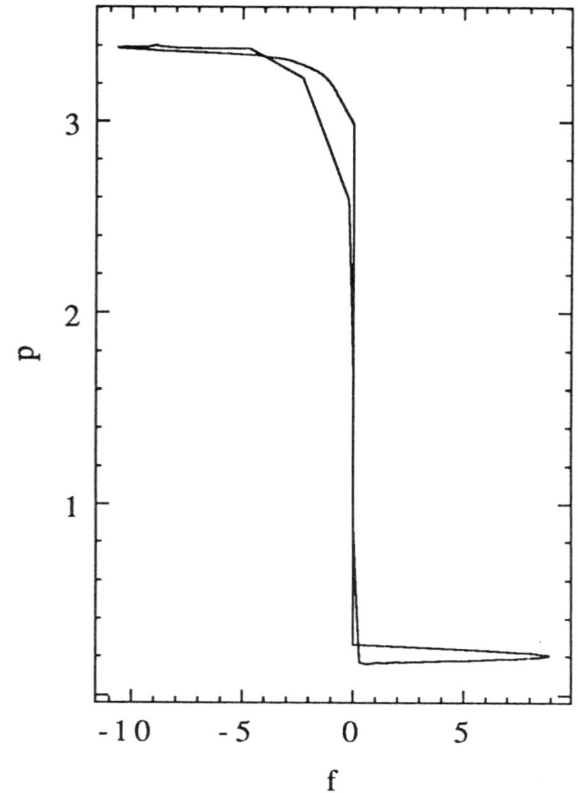

Fig. 8.9. (f, p)- graph for a butterfly catastrophe with ($a_2 = v$ and $a_4 = 2$)

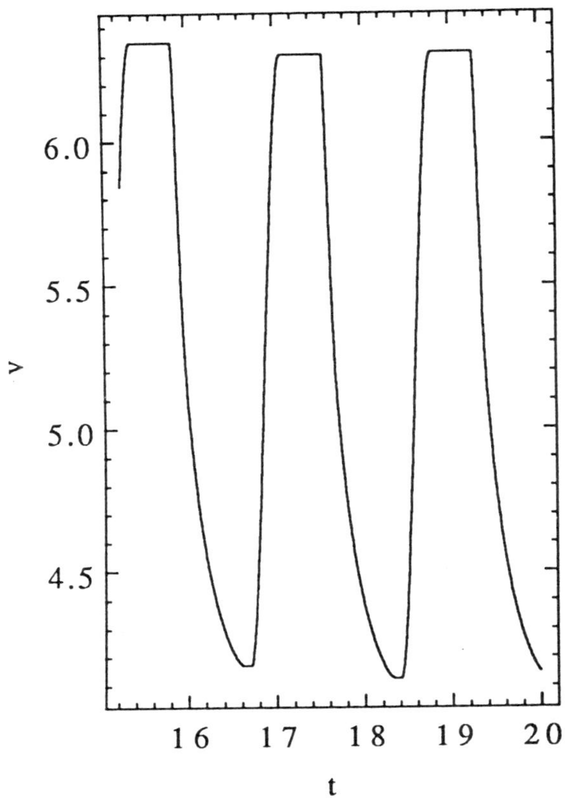

Fig. 8.10. (t, v)- graph for a butterfly catastrophe with ($a_2 = v$ and $a_4 = 2$)

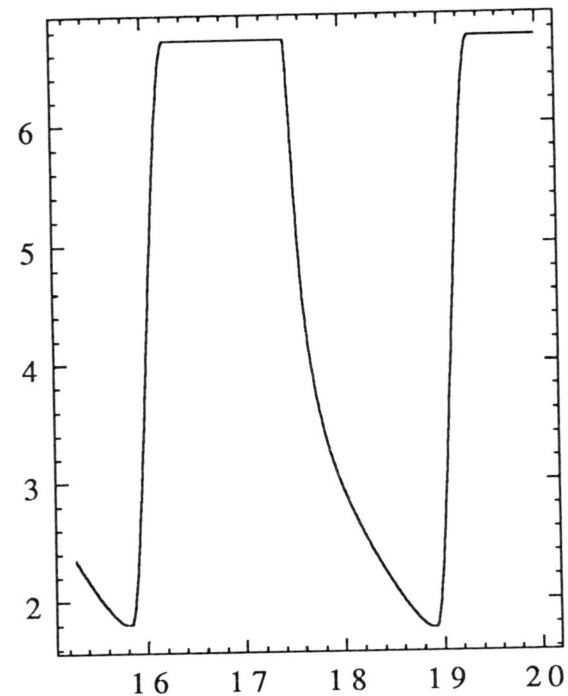

Fig. 8.11. (t, v)- graph for a butterfly catastrophe with ($a_2 = v$, $a_4 = P\text{aor}$ and $S = 14.5$).

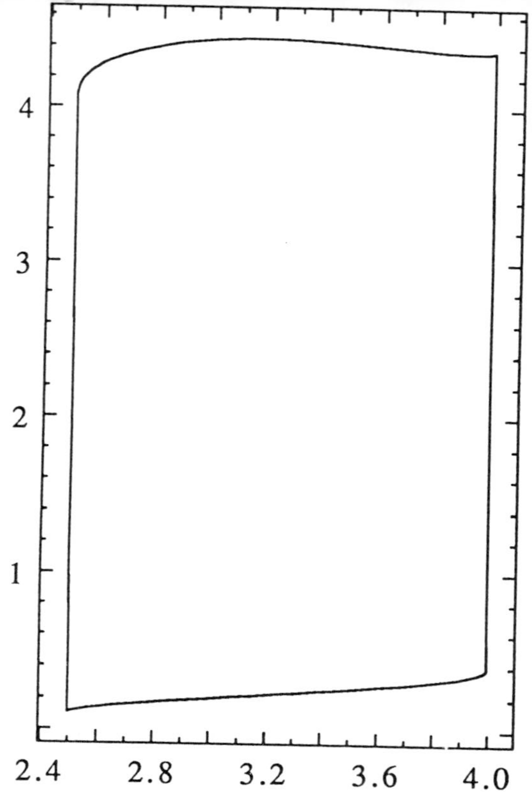

Fig. 8.12. (v, p)-graph for a butterfly catastrophe with ($a_2 = v$, $a_4 = P$aor and $S = 75$).

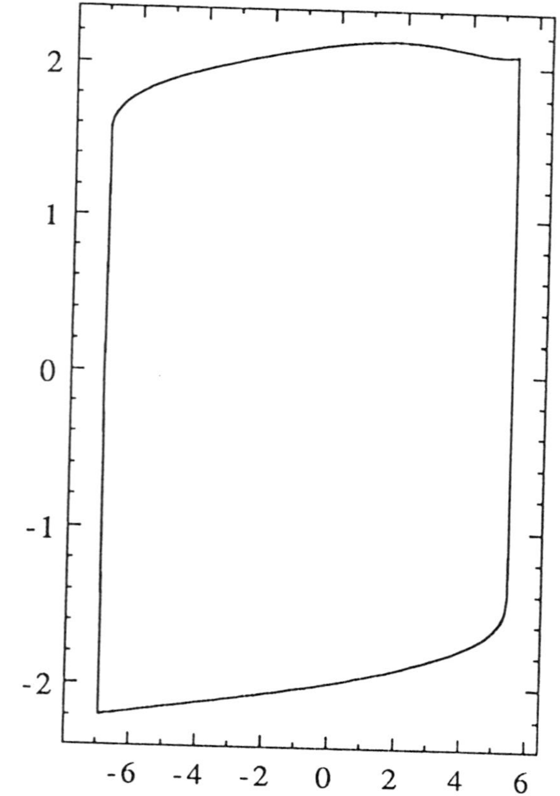

Fig. 8.13. (a_1, x)-plot for a butterfly catastrophe with ($a_2 = v$,$a_4 = P$aor and $S = 75$).

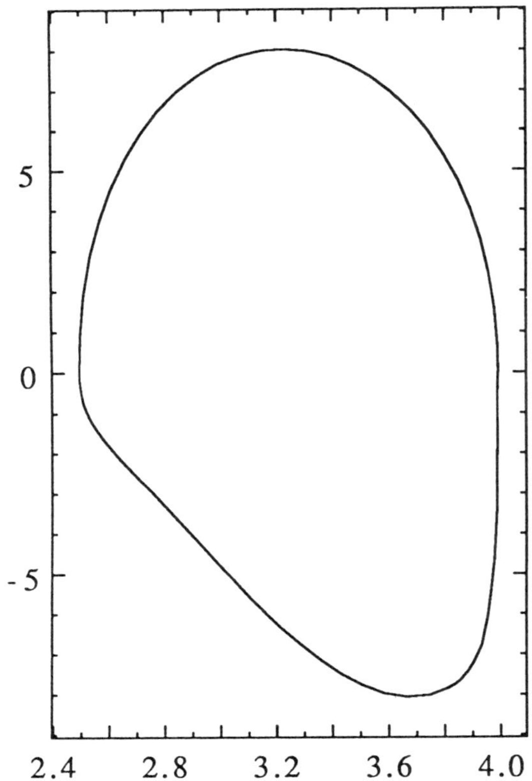

Fig. 8.14. (v, f)-graph for a butterfly catastrophe with ($a_2 = v$, $a_4 = P_{\text{aor}}$ and $S = 75$).

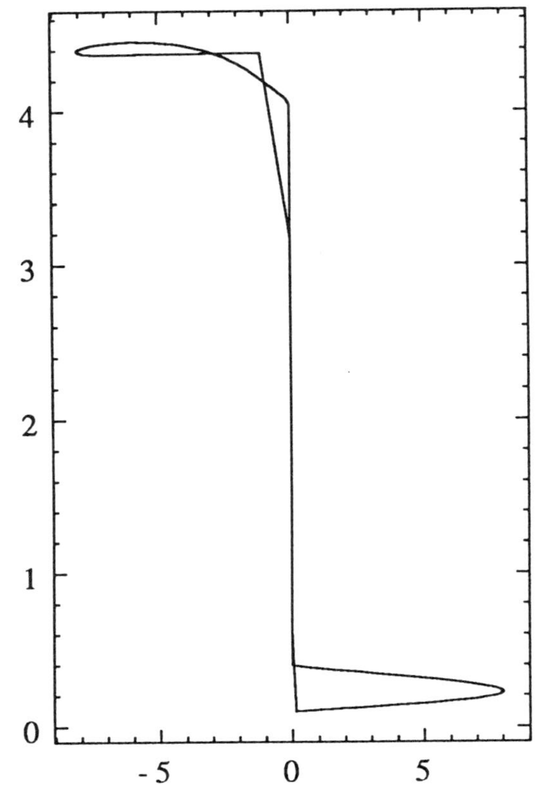

Fig. 8.15. (f, p)-graph for a butterfly catastrophe with ($a_2 = v$, $a_4 = P_{\text{aor}}$ and $S = 75$).

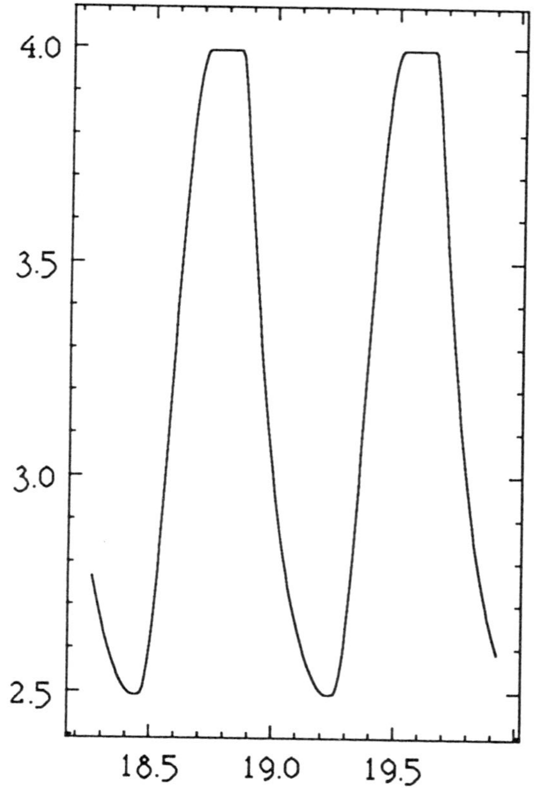

Fig. 8.16. (t, v)-graph for a butterfly catastrophe with ($a_2 = v$, $a_4 = P_{aor}$ and $S = 50$).

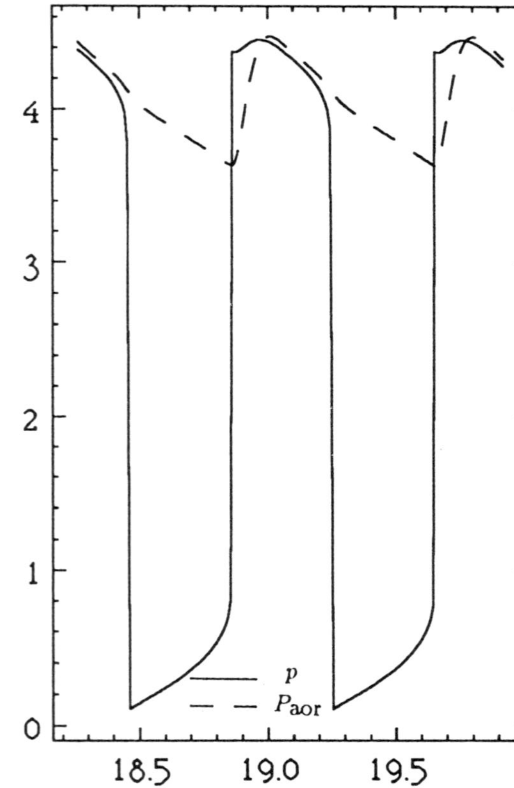

Fig. 8.17. (t, p)-graph for a butterfly catastrophe with ($a_2 = EDV$, $a_4 = P_{aor}$ and $S = 75$).

recorded graph (Fig. 5.17) during the pressure jumps as has already been discussed in Chapter 5. At the high pressure and low pressure levels the resemblance with the recorded graph is very good. The small loops in the recorded graph during the low pressure level are due to the atrium.

The simulated (t,v)-graph (Fig. 8.16) is approximately the recorded (t,v)- graph (Fig. 5.8). From EDV the volume decreases fast, but this decrease slows down till ESV is reached. When the AV-valves opens the volume increases rapidly and slows down again. In the recorded graph the supplementary increase is due to atrial contraction.

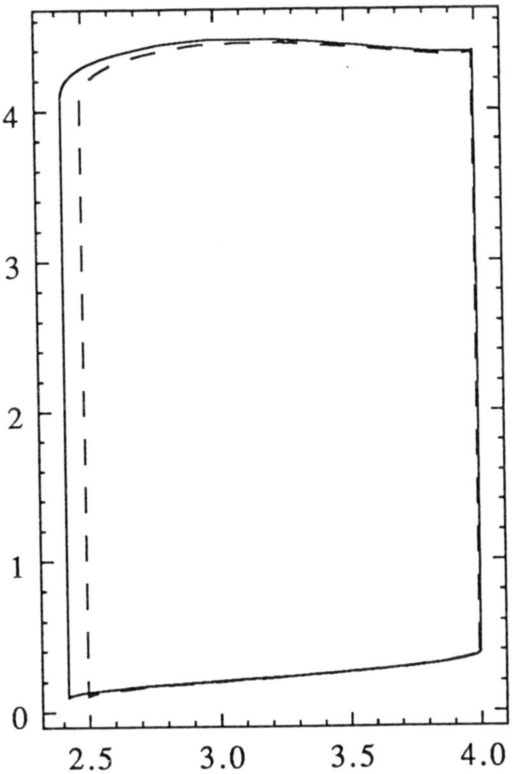

Fig. 8.18. (v,p)-graph for a butterfly catastrophe with ($a_2 = v$, $a_3 = 1$(—) or $a_3 = 0$(- - -), $a_4 = P_{\text{aor}}$ and $S = 75$).

The simulated (t,p)-graph(Fig. 8.17) resembles very well the recorded one (Fig. 5.11). From the low level the pressure jumps fastly, until the aortic valve opens. Then it increases till the PSP is gotten. From that moment, the pressure decreases. At first the decrease is slow, till the aortic valve closes and then jumps almost instantly to the low level. During diastole the pressure increases in both simulated and recorded graphs. The recorded graph also shows the atrial pulse, which blocks venous inflow and facilitates the atrial emptying in the ventricle. This is not present in the simulated one but the blocking of the venous flow is obtained by a steeper increase.

As shown from these graphs the cardio-vascular system can be simulated using a butterfly catastrophe model However, we have thus far no provisions that model adoptions when the subject changes its posture, activity etc.. These changes modify the contractility of the heart. Contractility is usually thought of as constant over one cycle, but changeable (by hormonal or nervous control) otherwise. The effect of a change in contractility is not uniquely determined. Two extreme situations are possible when contractility changes, namely: (i) upon constant PSP the SV increases, and (ii) at constant SV the PSP increases; all other situations lie in between these two, depending on the vascular situation. The extreme changes can be recuperated in the butterfly model by promoting the parameter a_3 to become the contractility parameter. Thus, the situation (i) can be seen on the simulation by comparing the (v, p)-graphs belonging to the values $a_3 = 0$ and $a_3 = 1$ as in Fig. 8.18. Here, the graph belonging to the former value is a full line while the latter value corresponds to the broken line. Indeed, we see that the PSP is the same in both situations, while the SV is larger for the largest value of a_3. Finally, the simulation of the situation (ii) where SV is constant and the PSP increases is obtained as follows. By a shortening of the period or by an increase of the peripherical resistance or the arterial capacitance, the pressure increases. This causes a decrease of the stroke volume, which can be adjusted by increasing the value of the parameter a_3.

Chapter 9 Recuperating the Frank and Starling Laws from the Model

Thus far we simulated the quasistatic behaviour of the cardiac ventricle as it integrates itself within the mammalian cardio-vascular system. The butterfly model that simulated the ventricle included inotropic as well as chronotropic parameters. We now show that the dynamic behaviour of the full cardio-vascular system system (without the atrium), and more in particular its autoregulation, can be simulated from the same butterfly model. Autoregulation of the cardio-vascular system has been known for about one hundred years, since Frank (1895) was the first to formulate the non-isometric part of what later were to be called the 'Frank-Starling laws'—which included non-isometric situations as well. Frank's Law says that

"the maximal ventricular pressure during an isometric contraction (i.c. when there is no outflow) is a monotonely increasing function of the preload".

Starling's Law is more sophisticated in that it draws the highly non-linear contractile (non-isometric) states of the heart muscle into the picture. In most medical textbooks it is formulated in a qualitative way by drawing the graph that plots ventricular performance vs. ventricular EDV; and then indicate how that graph changes its shape and raises or lowers its level as the contractile state of the heart muscle changes. However, there is also a more explicit graphical-numerical form of the Starling law entirely in the (p, v)-plane, as proposed by Sonnenblick [cf. Silbernagel, 1981]. The following rule of thumb sums up the case of constant contractility:

"For constant contractility and aortic pressure the stroke volume SV is an increasing function of the afterload EDV".

Below, we shall verify both laws from the present butterfly model as it has been coupled to the electric model for the vascular system. First of all, Frank derived his law from experiments with heart-lung preparations. In his experiment Frank studied ventricular inotropic behaviour by letting the ventricle pump against a series of increasingly high aortic pressures, so that in the end he could see what happened in the limiting no-outflow situation, i.e. when the ventricle pumps against a closed aortic valve. In our model this is tantamount to deleting the vascular part from Figure 5.1, i.e. by dropping the aortic pressure from the butterfly model. Thus, to simulate Frank's experiment, we are left with the question as to what to do with the contractility parameter a_3, and we set it equal to a constant, say zero. As a consequence, we are led to surmise that the (5,4)-cusp gives a good approximation of what happens in Frank's experiment. Indeed, even if we have shown in the previous chapter that in the normal, closed loop situation the simulated time graphs of heart rate, preload and afterload concur very well to both the time graphs and the time-free graphs on record respectively, it is incumbent upon us to show that in te border line case of a closed aortic valve we get a result similar to Frank's. That is to say, we have to show that the maximal ventricular pressure (PSP or peak systolic pressure) as it always lies in between ESV and EDV, is a monotonely increasing function of EDV. To prove this, first note that in the closed aortic valve case, the diastolic volume v = EDV = ESV; it is the only external control parameter left in the model. Equation 7.7.7 now says that the difference between the highest and lowest pressure d in the (5,4)-cusp case (i.e. the $a_3 = 0$ and $a_4 = 0$-case), is given by the formula

$$d = 2 * 1.10384\sqrt[4]{EDV}.$$

Figure 9.1 plots fifteen simulated values of the half-distance $d/2$ of d as a function of EDV and makes a curve fit as well (the full line (—)). Note that, in fact, PSP is equal to $d + c$, where c is a fixed resting pressure. The graph shows the monotony of PSP as required by Frank's law. This means that for the open (static) system our model accords to Frank's law. Note that, letting a_3 increase (resp. decrease), means shifting the curve upwards (resp. downward) while concomitantantly, its average radius of curvature decreases (resp. increases).

Frank's investigations were followed-up by Starling, who connected an artificial vascular system to the heart-lung preparation and kept cardiac rate, contractility and the mean aortic pressure constant in order to study the relationship between the preload and the stroke volume. A clear summary of these experiments was given by Sonnenblick in (Silbernagl (1981)) in a series of plots. In the presented pressure-volume plane (cf. Figure 9.2), three major curves are shown: the resting pressure-volume curve (—), the curve of the isotonic maxima (- + -) and the curve of the isovolumetric maxima (- - -).

The resting pressure-volume curve is obtained by filling the ventricle during diastole, without contraction. For different filling volumes the ventricular pressure is then measured and plotted. Note that this curve is, in fact, the same curve as that of Figure 9.1.

The curve of the isotonic maxima (- + -) is obtained by keeping the ventricular pressure constant during the ejection phase. For different filling volumes the minimal ventricular volume is then measured and plotted (horizontal arrows).

The curve of the isovolumetric maxima (- - -) is obtained by keeping the ventricular volume constant during the ejection phase. For different filling volumes the maximal ventricular pressure is then measured and plotted (vertical arrows).

The line joining the isometric maximum and the isotonic maximum related to one filling volume (EDV) is called the U-line. This line is the locus of the maximal pressure for different stroke volumes at that EDV. The maximal pressure (PSP) is indicated by P (notice that in reality the maximal pressure does not occur at minimal volume, so here there is simplification). Three other important points can be marked on the graph: The end of diastole (D), the opening of the ventricular valve (O), and the end of systole (S).

The pressure volume diagram can be schematically drawn by the quadrangle DOPS. The base of this quadrangle is the stroke volume SV. The ordinate of O is the aortal diastolic pressure. Using these diagrams the following three dynamic changes of the heart can be represented and made visual with a hint to quantification:

1. the effects of changes in bloodvolume (cf. Figure 9.3),
2. the effects of changes in blood pressure (cf. Figure 9.4) and
3. the effects of changes in contractility (cf. Figure 9.5a and b).

As for (1), a greater venous blood volume means a larger EDV, i.e. a shift of the point D to the right (D′). This causes a new U-curve (U′). The point P changes to P′ on the new U-curve with approximately the same PSP. The related ESV and SV increase as well, with S shifting to S′. Conversely, when the blood volume reduces, then the EDV,

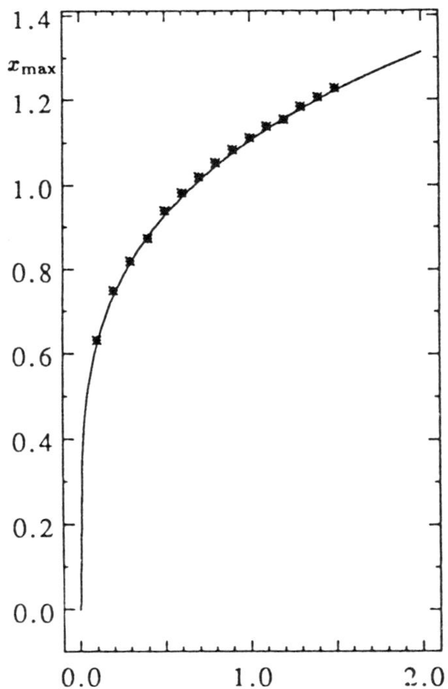

Fig. 9.1. The half distance $d/2$ between the lowest and highest pressure is plotted as a function of EDV from equation 7.7.7; the dots show simulated values.

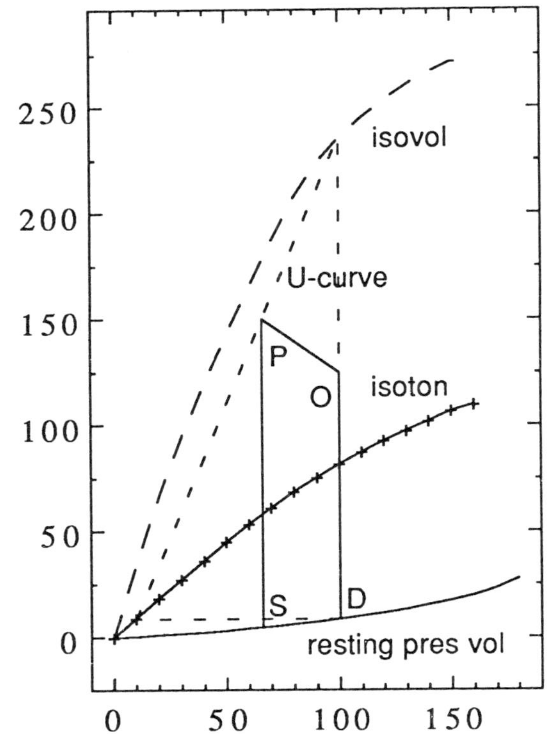

Fig. 9.2. The pressure-volume curves after(Silbernagl (1981)). The resting pressure volume curve (—), the curve of the isotonic maxima (- - -) and the curve of the isovolumetric maxima (- + -). The line joining the isometric maximum and the isotonic maximum related to one filling volume (EDV) is called the U-line.

ESV and SV decrease while the maximal pressure (PSP) remains constant as well. This is in accordance with the Starling Law. Starling found that the cardiac output increased as the preload increased. Since the cardiac output is equal to SV/T, where SV is the stroke volume and T the heart rate, and moreover, since the preload increases with EDV, the above rule of thumb is part of the Sonnenblick representation.

As for (2), increasing the blood pressure means shifting O to O″ (Figure 9.4). The immediate result is an increase of PSP and the point P moving on the same U-line to P″, causing an increase of ESV while SV is reduced at least for this beat. If the filling rate of the ventricle is maintained, then the EDV will increase and D will slide over the resting volume curve to the right. This situation will continue until the SV is about the same as the initial SV. Comparing the initial with the final quadrangle D"O"P"S, we observe that the maximal pressure PSP, as well as EDV and ESV, increase while SV remains constant.

As for (3), an increase in contractility causes the curve of the isovolumetric maxima (---) and consequently the U-line (U‴) to shift upwards. Now, the graph may develop in any way lying in between two extreme possibilities, as Sonnenblick in (Silbernagl (1981)) makes explicit in terms of his diagrams. The first extreme is that EDV, ESV and SV remain constant (Figure 9.5a). In that case PSP will increase or the PSP remains constant. The other extreme possibility is that ESV may decrease at constant EDV, so that consequently SV increases (Figure 9.5b, quadrangle DOP⁗S⁗).

We have been simulating the three dynamic changes in our butterfly model.

1. Indeed, the effect of an increase in blood volume resulted in ann increase of EDV, ESV and SV. In Figure 9.6 the (v, p)-graph of this simulation is shown. The broken line (---) represents the situation with the largest volume while the full line (—) represents the situation where the volume is 10 pct less, showing an increase of blood pressure. This figure compares to Figure 9.3 by Sonnenblick. It shows that our simulation accords with the Starling Law.
2. Figure 9.7 shows that a reduction in arterial compliance results in increased blood pressure. The full line graph represents the low blood pressure situation, while the broken line represents the resulting increased pressure. SV remains the same while PSP increases. Again, the simulation matches with the description by Sonnenblick (Figure 9.4).
3. In Chapter 8 (Figure 8.18) we have already shown that if myocardial contractility is changed by having a_3 increase from zero to one, then SV increases while EDV and PSP remains the same. This accords with Sonnenblick's representation as well (Figure 9.5b). The simulation corresponding to Figure 9.5a also accords with Sonnenblick's representation. It will be published in a separate paper that also contains simulations of some major heart diseases.

We may conclude that the butterfly model simulates not only the static process but also the dynamic changes of the real heart.

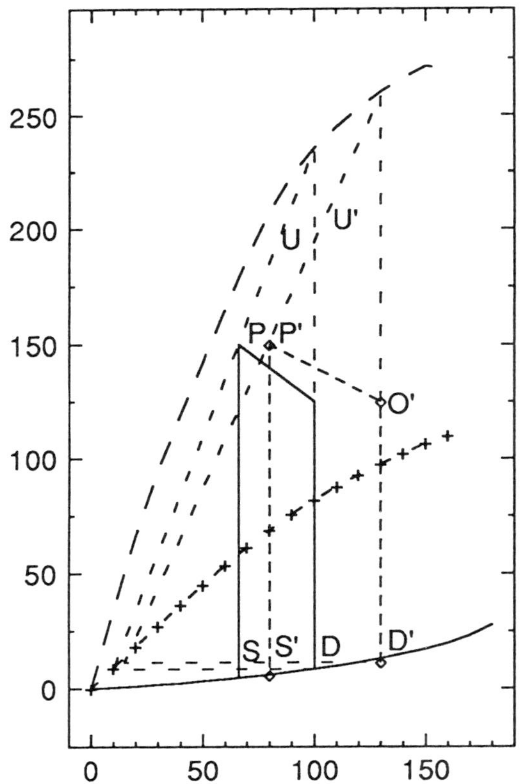

Fig. 9.3. The pressure-volume curves after(Silbernagl (1981)) of the effect of an increase in bloodvolume, smallest volume (—), largest volume(- - -)

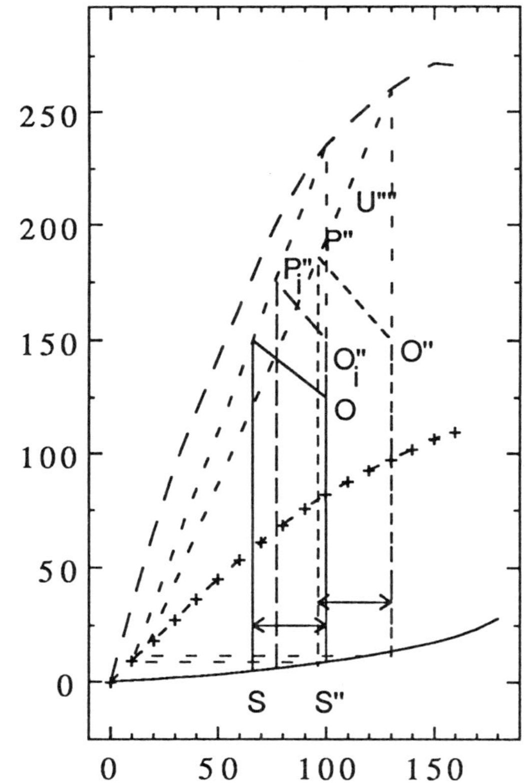

Fig. 9.4. The pressure-volume curves after(Silbernagl (1981)] of the effect of increase in blood pressure, smallest pressure (—), largest pressure(- - -)

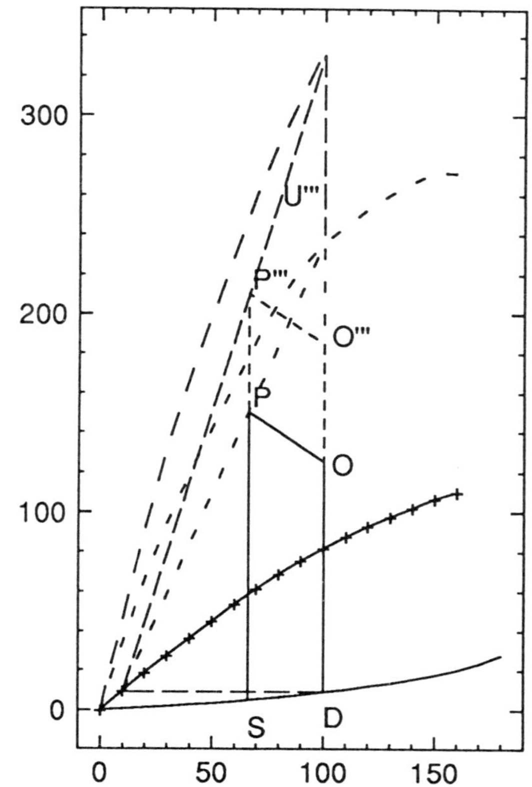

Fig. 9.5a. The pressure-volume curves after (Silbernagl (1981)) of the effect of an increase in contractility, where EDV, ESV and SV remain constant, smallest contractility (—), largest contractility(- - -)

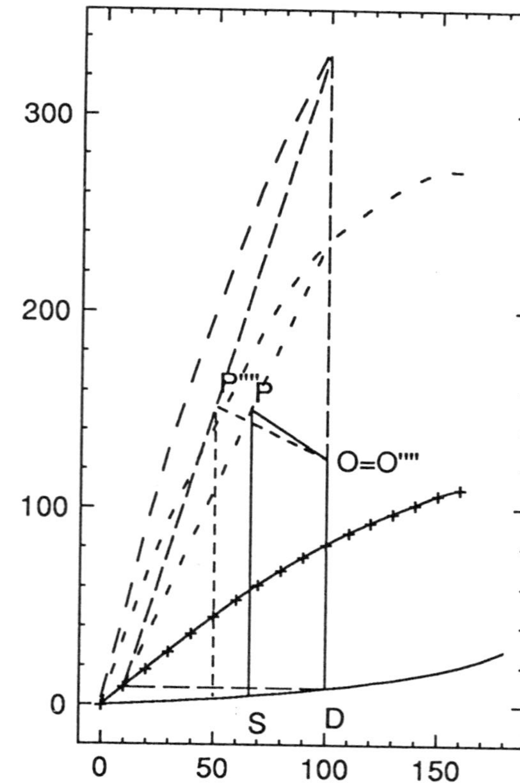

Fig. 9.5b. The pressure-volume curves after (Silbernagl (1981)) of the effect of an increase in contractility, where PSP remains constant, smallest contractility (—), largest contractility(- - -)

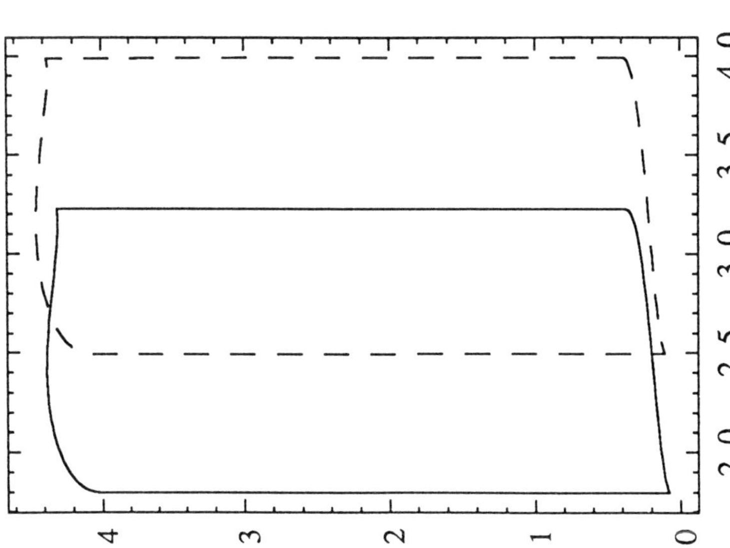

Fig. 9.6. The pressure-volume curves of the simulation using the butterfly catastrphe model on the effect of a change in blood-volume, smallest volume (—), largest volume(- - -)

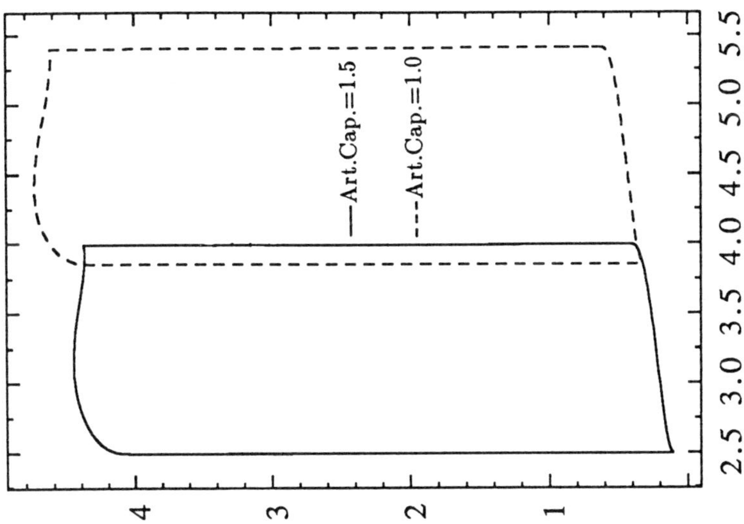

Fig. 9.7. The pressure-volume curves of the simulation using the butterfly catastrphe model on the effect of a change in blood pressure, smallest pressure (—), largest pressure(- - -)

References

E.O. Attinger and A. Anné (1966), Simulation of the cardio-vascular system, *Ann. New York Acad. Sci.* 128, 810-829.

A.A. Bove and P.R.Lynch (1970), Radiographic determination of force-velocity-length relationship in the intact dog heart, *J. Appl. Physiol.*, Vol 29, No 6.

A.F. Bertelsen, L. Walloë and Arve Kvalheum (1986), Blood flow simulations in a cast of the aortic bifurcation, Preprint Series ISBN 82-553-0621-8 Appl.Math.

J.R. Brobeck (1979), Best & Taylor's Physiological Basis of Medical Practice, 10th Ed., J.R. Brobeck Editor, The Williams & Wilkins Co, Baltimore.

W.F. Ganong (1973), Review of Medical Physiology, 6th Ed., Lange Medical Publications, Los Altos.

J.L. Janssens, E.R. Raman, and V.J. Vanhuyse (1989), A pressure-controlled fluid flow generator for arterial blood-flow simulation, *J. Med. Eng. & Tech.*, 13, 104-108.

J.L. Janssens and E.R. Raman (1991), An electrical analog for a pressure-controlled, fluid flow generator for arterial blood-flow simulations, *J. Med. Eng. & Tech.*, 15, 21-25.

W. Kuyk (1986), A Catastrophe Theoretic Model of Heat Regulation in Homeothermic Animals, *Bull. math. Biol.*, 46, 81-102

W. Kuyk (1986) The Human Thermostat, A Catastrophe Type Controlling Biological Processes, in: G.C. Vansteenkiste ec. Editors. Proceedings 2nd European Simulation Congress, Antwerp 1986, 713-719.

W. Kuyk (1988), Servocontrol of the Heart Rate in Homeothermic Animals: On Recognizing Catastrophy Models in Biology, *Bull. math. Biol.*, 48,107-124.

W. Kuyk and L. Smits (1990), On the Geometries of the Universal Unfoldings of X^k, *Acta Appl.Math.*, 19, 77-86.

T. Poston and I. Stewart (1977), Catastrophe Theory and its Applications, Ed. Pitman, London.

S. Silbernagl and A. Despopoulos (1981), Sesam Atlas van de Fysiologie, Ed. Bosch en Keuning nv, Baarn.

L. Smits and W. Kuyk (1990), Winding number and the number of real zeros of a function, to appear in *Proc. Am.Math.Soc.*, 11, No 4, April 1992, 981-987.

M.F. Snyder and V.C. Rideout (1969), Computer Simulation Studies of the Venous Circulation, *IEEE Trans. Bio.-Med. Eng.*, BME-16, 325-334.

M.F. Snyder, V.C. Rideout and R.J. Hillestad (1968), Computer modeling of the human systemic arterial tree, *J. Biomech.*,1, 341-353.

J.J. Stoker (1950), Nonlinear Vibrations, vol. II, Interscience Publishers, New York.

E.C. Zeeman (1977), Catastrophe Theory, Collected papers. Ed. Addison-Wesley Publishing Cy., London.

Figure References

A.A. Bove and P.R.Lynch (1970), Figure:4.1

J.R. Brobeck (1979), Figures: 1.1-2, 1.4, 4.6

W.F. Ganong (1973), Figures: 1.3, 1.5-7,5.7,5.9,5.11

T. Poston and I. Stewart (1977), Figures: 2.1-7, 3.4, 7.1-2, 7.4

S. Silbernagl (1981), Figure: 9.4

M.F. Snyder and V.C. Rideout (1969), Figure: 1.9

M.F. Snyder, V.C. Rideout and R.J. Hillestad (1968), Figure: 1.8

J.J. Stoker (1950), Figures: 3.2-3, 3.5

E.C. Zeeman (1977), Figure: 3.1

Index

About the Authors

Born in 1957 in Beveren, Belgium, **Jan L Janssens** studied Mathematics and Physics at the University of Antwerpen, from which he received an MSc, 1982. In 1986 he completed a second MSc at the Free University of Brussels in Biomedical and Clinical Engineering Techniques. Working and publishing on arterial fluid transport problems, he also worked on mathematical models for the cardiovascular system. In 1990 he received a DSc at the University of Antwerpen following work in the subject of this monograph. He is presently continuing research on the general mathematics of transport problems.

Born in 1934 in Amsterdam, **Willem Kuyk** studied Mathematics and Physics at the Free University of Amsterdam, where he received his DSc degree in 1960 in Galois Theory and Number Theory. He was a Fellow of the Mathematical Centre, Amsterdam, from 1961-1963, and a lecturer at the University of Amsterdam during 1962-1963. In 1962 he was awarded a prize from the Dutch Mathematical Society, *Het Wiskundig Genootschap*. He held a Postdoctoral Fellowship of the National Research Council of Canada in 1963, and an associate professorship at McGill University (Montreal) during 1964-1970. He became full professor at the University of Antwerpen in 1969, where he continued research in Number Theory and Algebra. Further research in catastrophe theoretical models in Physiology led to his present area of interest, which includes models for heat regulation in homeothermic animals, the matching simulation method (which is applied in the present volume), and a series of papers in Quantum Neural Networks. He has edited and co-edited several books on Number Theory, and authored one book on the Foundations of Mathematics. He has been a visiting member of the Institute for Advanced Study in Princeton NJ, USA, in 1968 and 1974.